SKETCHUP FOR CIVIL EN
AND HEAVY CONSTRUCTION

About the Author

Vladimir Simonovski, M.Sc., is a structural engineer and co-founder of VLS Engineering, LLC., a structural consulting company comprised of highly skilled engineers, designers and forward-thinkers.

Prior to co-starting his own company, Vladimir gathered extensive design and construction knowledge and experience by working on high profile infrastructure projects around the U.S. On some projects, his experience was gathered by being a part of the design team and on other projects Vladimir made his presence know in the field, by directly getting involved in construction, as part of the construction team. This balanced mix of experience, coupled with a strong work ethic, gave Vladimir the outlet for his interest in reinventing how problems are viewed and finding solutions by utilizing technologies such as 3D modeling, virtual reality, 3D printing, scanning, implementation of new materials and etc.

His enthusiasm can be seen from the numerous three-dimensional modeling manuals that he has prepared over the years. He has conducted training for construction companies on the effective use of the SketchUp application and how it can fill the effectiveness gap where other CAD applications cannot. During these training sessions, a special emphasis is given to the idea of Information Modeling and the proper organization in the heavy civil industry. Additionally, his 3D modeling work was published by the American Society of Civil Engineers - ASCE, under the practice periodicals on structural design and construction. The paper was written based on a real-life beam erection scenario, in which Vladimir was heavily involved and where 3D modeling was used extensively to plan and test multiple critical girder picks.

Besides his everyday engineering obligations to current projects, Vladimir and his team are also involved as technical consultants for the development of new construction materials and his personal interest, construction beyond our planet.

SKETCHUP FOR CIVIL ENGINEERING AND HEAVY CONSTRUCTION

Modeling Workflow and Problem Solving for Design and Construction

VLADIMIR F. SIMONOVSKI

New York   Chicago   San Francisco
Athens   London   Madrid
Mexico City   Milan   New Delhi
Singapore   Sydney   Toronto

Library of Congress Control Number: 2021935623

McGraw Hill Education books are available at special quantity discounts to use as premiums and sales promotions, or for use in corporate training programs. To contact a representative please visit the Contact Us page at www.mhprofessional.com.

SketchUp for Civil Engineering and Heavy Construction: Modeling Workflow and Problem Solving for Design and Construction

1 2 3 4 5 6 7 8 9   DSS   26 25 24 23 22 21

ISBN   978-1-260-46038-4
MHID   1-260-46038-X

The pages within this book were printed on acid-free paper.

Sponsoring Editor	**Proofreader**
Ania Levinson	Wendy Jo Dymond
Editorial Supervisor	**Indexer**
Donna M. Martone	Vladimir F. Simonovski
Acquisitions Coordinator	**Production Supervisor**
Elizabeth Houde	Pamela A. Pelton
Project Manager	**Composition**
Jyoti Shaw, MPS Limited	MPS Limited
Copy Editor	**Art Director, Cover**
Bindu Singh, MPS Limited.	Jeff Weeks

For my father Filip Simonovski,
the greatest journalist and writer that I ever knew
and for my family and especially my son Ordence!

Contents

vii

Introduction

SketchUp for Civil Engineering and the Heavy Construction Industry represents a collection of topics, workflows, and examples gathered throughout my career as a structural engineer. I had two goals in my mind when I was writing this book. The main goal of this book is to introduce the concept of **Information Modeling and Organization**. The core idea behind this concept is to create three-dimensional drawing at the infancy stages of a construction project, and use those same drawings, through smart organization, for design, estimation, management, and construction purposes. The **Information Modeling and Organization** concept is explained in detail in **Chapter 4** of this book. My second goal was to create a SketchUp how-to manual written by a civil engineer that other civil engineers will be able to relate to, based on their day-to-day activities.

The overall need for this SketchUp book comes from advances in design and construction technology. As is the case with other construction industries, the heavy civil industry advances rapidly forward in technology and design. This trend can be seen on a daily basis by the increase in the number of highly complex structures that are currently or will be designed and scheduled for construction. Intelligent bridges, airports, tunnels, water control structures, rail-roads, and roadway lines, all fall under this push into the future of design and construction.

From my own work experience, this increase in design complexity also brings an increase in construction precision. Small tolerances coupled with highly complex details decrease the efficiency of standard two-dimensional drawings to convey all the necessary information to the engineers and constructors working in the field. Intelligent three-dimensional drawings used in conjunction with tablets and smart phones become the new norm on projects. SketchUp can fill the gap where other, more complex CAD applications simply cannot. With its overall characteristic simplicity to learn, use, integrate and the availability of a wide array of extension applications, SketchUp is the right three-dimensional computer drafting application to take two-dimensional drawings to the next level. My overall hope is that this book becomes *Your* how-to manual that it was envisioned to be, and will help you on projects, for years to come.

For additional help, please visit us at **www.vlsengineering.com** and check out the INFORMATION MODELING or the BLOG page for additional information on various topics and modeling solutions.

Introduction to SketchUp

The set of topics that are covered in **Chapter 1** of this book are intended to give the reader a sufficient understanding or base knowledge as a prerequisite in order to follow and complete the examples that are covered in later chapters. The aim of *SketchUp for Civil Engineering and the Heavy Construction Industry* is to be equally accessible and usable by professionals with basic and advanced skill sets in SketchUp.

With this in mind, **Chapter 1** focuses on explaining the basic operational concepts of SketchUp and further explaining the available tools, settings, and options. Although this may be considered somewhat of a repeat and not needed for professionals who possess an advanced skill set in SketchUp, it is encouraged that you follow what is covered in **Chapter 1**, because you never know what you will learn or how your knowledge of a topic will increase.

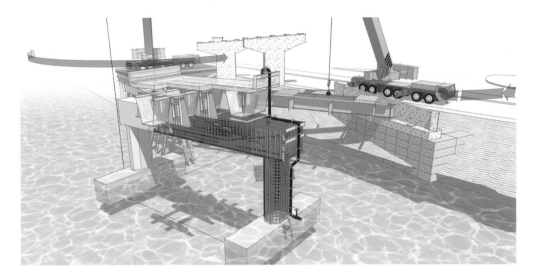

The SketchUp Concept

Before the review of SketchUp tools, settings, and options can begin, you need to explore and understand several basic operations and functions of SketchUp. Understanding these key concepts of SketchUp can make a big difference in your three-dimensional modeling experience and how you use the application on a day-to-day basis. There are a total of four main operational properties of how the SketchUp application functions: **Edges and Surfaces**, **Multiple Edges and Surfaces on a Same Plane**, **Representation of Curves**, and **Connection (Stickiness) of Surfaces**.

Edges and Surfaces

SketchUp is a three-dimensional computer drafting application that is based upon a concept of edges (lines) and surfaces (**Fig. 1.1**). A minimum of three edges, which are coplanar (lie in the same plane), are needed in order to make the most basic building block in SketchUp, which is a triangular surface.

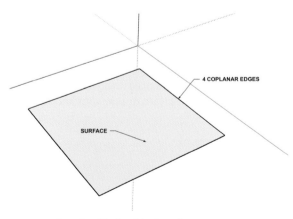

Figure 1.1   Representation of an **Edge (Line)** and **Surface** plane.

Multiple Edges and Surfaces on a Same Plane

When two or more surfaces come into contact at the same plane (i.e., they occupy the same space), any portion of the edge and surface that overlap becomes a new entity (**Fig. 1.2**). In other words, the union of these two surfaces will create a separate new surface. Any edge that is overlapped by another edge will split that edge into two sections (**Fig. 1.2**).

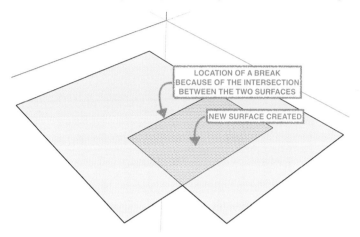

Figure 1.2   Break in a line and surface because of an intersection of two surfaces located on the same plane.

Representation of Curves

SketchUp represents circles, arcs (any type of curve in general), as a series of interconnected edges (**Fig. 1.3**).

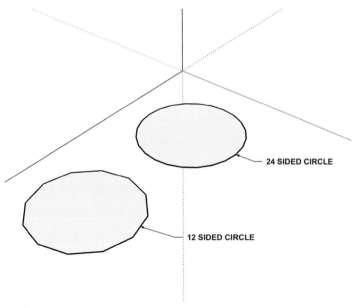

Figure 1.3   Visual difference between 24-sided circle and 12-sided circle.

The smoothness for all types of curves can be modified by changing the number of interconnected edges present in a specific curve—this property is discussed in more detail later in this chapter with the circle and arc tools.

Connection (Stickiness) of Surfaces

When two or more adjoining surfaces come into contact with each other (**Fig. 1.4a**), they join together and create an additional, separate surface (**Fig. 1.4b**). If you try to move one of the intersected surfaces or the newly created surface from the intersection, you will notice that it affect the overall geometry of your drawing, see Fig. 1.5.

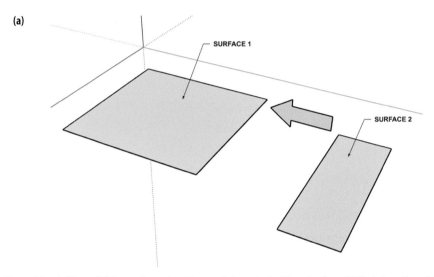

(a)

SURFACE 1

SURFACE 2

Figure 1.4   (a) Two adjoining surfaces about to come into contact with each other. (b) The interaction of the two surfaces creates a new surface in the intersected area of the two.

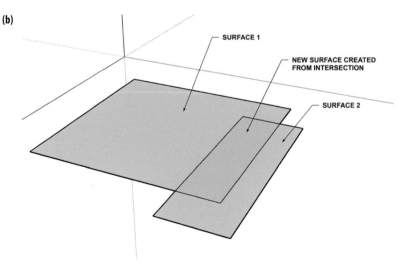

(b)

SURFACE 1

NEW SURFACE CREATED FROM INTERSECTION

SURFACE 2

Figure 1.4   (Continued)

This operational property of SketchUp is very important to understand since it will affect how a collection of surfaces and/or edges will react when they are modified (**Fig. 1.5**). This property is no longer valid if the surfaces are either grouped or turned into components.

Now that you understand these basic concepts in SketchUp, you can start exploring everything SketchUp has to offer for the heavy civil engineering and heavy construction industries.

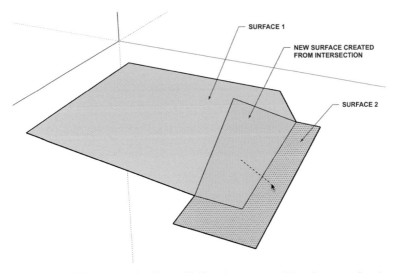

Figure 1.5   Moving one of the intersected surfaces will affect the geometry of the other two surfaces by modifying it. (This property is very important to understand since it can affect your three-dimensional model when you least expect it if your objects are not grouped or made into components.)

Welcome Screen and Templates

The **Welcome to SketchUp** window represents the first set of options that are presented by SketchUp (**Fig. 1.6**) when the application is started.

Most of the options on the welcome screen are self-explanatory and therefore are only briefly mentioned.

The following list represents all the options present at the welcome screen (**Fig. 1.6**):

1. Allows you to manage your profile, subscription, colaboration, etc.
2. Location where your favorite templete is shown in addtion to the mostly used templates. Templates will be discussed later in this chapter.
3. Shows all the options connected to files and the use of files.
4. Location where you can receive additional training, that be through videos, forums and etc.
5. Button that allows you to open a specific file.
6. Area where the recently used files are shown.
7. Allows you to specify how items are shown.
8. Area where additional information is shown in regards to your license.

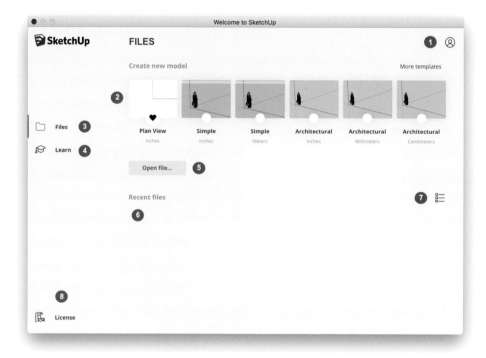

Figure 1.6 Welcome to SketchUp screen—the first set of options available to the user.

SketchUp Templates

SketchUp provides a total of 19 different templates to be used and modified as needed by the modeler. The full list of templates can be accessed by selecting **File** from the menu and then selecting **New From Template…** (**Fig. 1.7**).

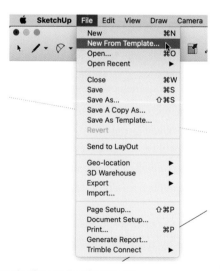

Figure 1.7 Menu access to the Choose a Template window.

When you complete the selection, a secondary window called **Choose a Template** will open, where all 19 templates are shown (**Fig. 1.8**).

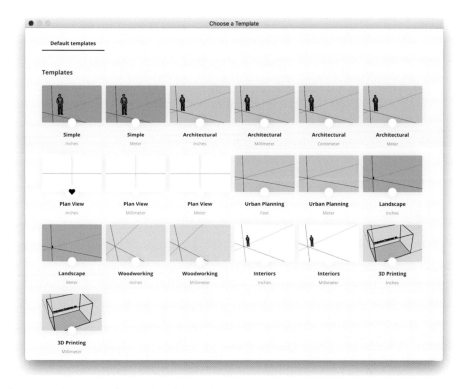

Figure 1.8   Choose a Template window, showing the 19 different default template options available for use.

The 19 different templates can be divided into 8 individual categories:

1. Simple
2. Architectural
3. Plan View
4. Urban Planning
5. Landscape
6. Woodworking
7. Interiors
8. 3D Printing

Each of these categories can be further subdivided by unit types (S.I. or U.S.C.) and unit scale (inches, feet, millimeters, meter, etc.). It is also very important to notice that besides the measurement differences, templates also differ by style (differences in back-ground masks, edge differences, view style differences, etc.).

Modeling Tip

This book will mostly use the Plan View (inches) template. In order to set this or any other template of your choosing as a default template, in the **Choose a Template** second-ary window click the white dot in the center of the previewed template. When the white dot is selected, it will place a heart icon inside the dot, which will signify that this is now your default template (**Fig. 1.9**). The default status can be changed by selecting another template and placing the heart icon as described earlier.

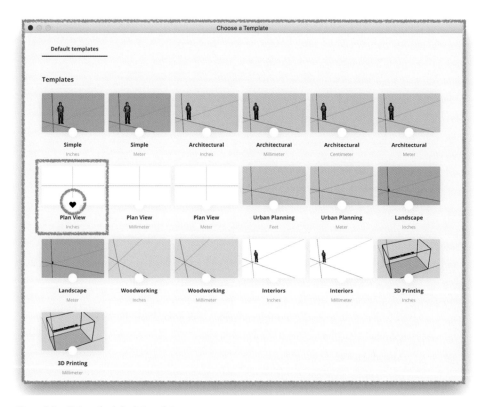

Figure 1.9   Setup of a default template.

SketchUp Application Window

Before our attention is focused on the different varieties of tools and toolboxes, it is very important to review the main SketchUp application window.

Main SketchUp Application Window

In the next part of this chapter, a basic overview is provided for all the sections in the SketchUp application window that are important to be fully comprehend. The section overview can be of tremendous help for your overall workflow and day-to-day work activities. Curiosity is a very important trait of a maker, and therefore, some room for self-exploration is left for the other options that SketchUp offers but are not reviewed in this part of the chapter.

For any of you who had the pleasure of hand drawing plans, think of the main application window in SketchUp as your drafting desk. You had your paper area, which represents your main modeling window in SketchUp; the different tins full of mechanical pencils, pens, erasers, color pencils, protractors, and so on represent your toolbar; and the small pockets of notes taped all around your desk represent your main menu bar. **Figure 1.10** shows my current setup of the main application window. This is my personal setup that functions very well with my workflow process; you are encouraged to find what works best for you (any way you would like to organize your drafting desk!).

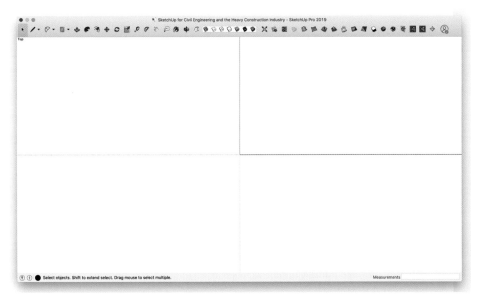

Figure 1.10   **Main Application** window and tools setup.

The main application window can be separated into six parts (**Fig. 1.11**):

1. Menu Bar
2. Tool Bar
3. Modeling Window
4. Measurements Input Box
5. Different Dialog Boxes
6. Third-party Extensions

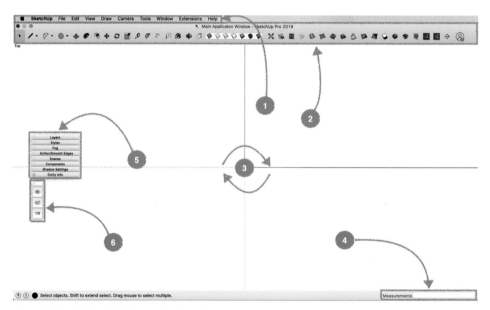

Figure 1.11   **Main Application** window breakdown.

Of these, parts 1 through 4 represent main application sections and are explained further in this chapter. The dialog boxes shown in Part 5 elements are explained throughout the book as they are being used in examples, and part 6 is touched upon in **Chapter 2—SketchUp and Resources**.

Menu Bar

Window, **Extensions**, **Model Info,** and **Preferences** are some of the more important menu options located under the menu bar. The **Entity Info**, **Components**, **Styles**, and **Scenes** dialog boxes (**Fig. 1.12**) can be found under the **Window** menu. These four dialog boxes are heavily used throughout the book.

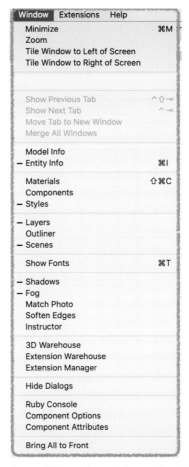

Figure 1.12   The **Window** menu is one of the more important menus in SketchUp, since it contains all the necessary inspection wondows and dialog boxes which are used on daily basis

The **Extensions** menu shows all of the third-party extensions that are installed under SketchUp (**Fig. 1.13**).

Figure 1.13   The **Extensions** menu.

If the installed extension does not show an icon in the toolbar or on the modeling window, this is the place where you can access it. Finally, the **Model Info** and the **Preferences** dialog boxes are used to adjust the SketchUp preferences per your requirements.

Modeling Tip

Some of the same tools and options can be accessed either from the **menu bar** or the **toolbar**.

The **Preferences** can be accessed by clicking on the **SketchUp** menu and selecting **Preferences** (**Fig. 1.14**).

Figure 1.14   Menu access to the **SketchUp Preferences** window.

When the selection is made, the **SketchUp Preferences** dialog box will appear (**Fig. 1.15**). The two key tabs **General** and **OpenGL** will now be discussed.

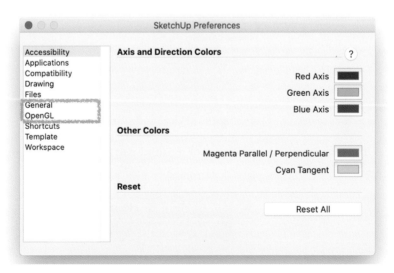

Figure 1.15   **SketchUp Preferences** window.

Under the **General** tab you will find the **Saving** settings. The default **Auto-Save** time is set to 5 minutes, but this can be changed to a shorter period, which is recommended especially if you work with large three-dimensional models that demand high computer power.

The second key tab that will be explained further is the **OpenGL** tab. Under the **OpenGL** tab, you will find the **OpenGL Settings**. The settings will vary from computer model to computer model and the make and model of the GPU—Graphic Processing Unit. **Figure 1.16** shows an example of standard setting under the **OpenGL** tab, You can change and manipulate these settings on a regular basis in relationship to the task that you are preforming and the three-dimensional model you are making.

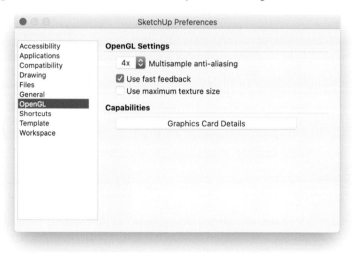

Figure 1.16   **SketchUp Preferences** window, **OpenGL** option activated.

The **Multisample anti-aliasing** drop-down menu allows you to change the quality of the drawing. As you choose a higher number before the "X" letter (**Fig. 1.17**) the less boxy the lines will appear on your drawing. Keep in mind that choosing a higher number from

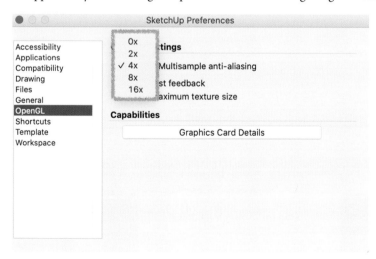

Figure 1.17   **SketchUp Preferences** window, **Multisample anti-aliasing** drop-down menu activated.

the drop-down menu also puts a higher demand on your computer system, which has a tendency to slow down the response time of SketchUp. It is recommended to change these settings when the three-dimensional model is completed and the time has come to work on the presentation portion of your project.

The **Use Fast Feedback** option is dependent and automatically enabled if your graphic card can support it. The **Use Fast Feedback** option is especially useful when working with large SketchUp models, which exert higher demands on your system, by improving the response time.

Similar to the **Multisample anti-aliasing**, it is recommended to use the **Maximum Texture Size** option when the three-dimensional model is completed and you begin working on the presentation portion of your project. As the option implies, enabling the **Maximum Texture Size** option will allow SketchUp to render texture images in resolutions larger than 1024 × 1024. If this option is not enabled, SketchUp will render texture images to a maximum resolution of 1024 × 1024.

The **Model Info** can be accessed by clicking on the **Window** menu and selecting **Model Info**. When the selection is made, the **Model Info** inspector window will appear (**Fig. 1.18**).

Most of the tabs that are present under **Model Info** are self-explanatory in nature and therefore do not need a deeper explanation. That being said, three tabs that are an important part of the **Model Info** are **Animation**, **Geo-location,** and **Units**.

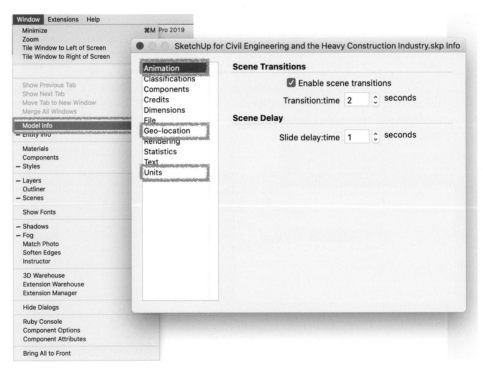

Figure 1.18   Menu access to the **Model Info** window.

The **Animation** tab allows you to modify the scene transition and delay time (**Fig. 1.19**), which is important when animations are created.

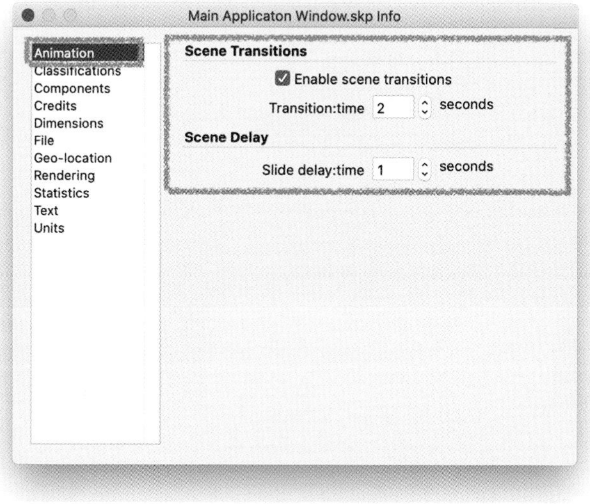

Figure 1.19   **Model Info** window shown with the **Animation** option activated.

The **Geo-location** tab allows you to geo-reference your drawing (**Fig. 1.20**), which helps you in two ways: (1) when creating terrain models by importing satellite images and data, which is discussed in more detail in **Chapter 9—Site Modeling and Use of Site Creation Tools,** and (2) when you need an accurate representation of shadows and shading of your three-dimensional model based on the location, which is discussed in more detail under. **Chapter 12-Introduction to Scenes, Section Cuts, Shadows, and Fog.**

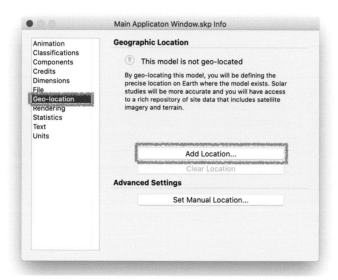

Figure 1.20   **Model Info** window shown with the **Geo-location** option activated.

The **Units** tab (**Fig. 1.21**) allows you to specify the format (Architectural, Decimal, Engineering, and Fractional) and the precision of the display units. Since all the examples are created with the U.S. customary system, the units are set to Architectural and will be displayed as 00′-00″, with a precision of 1/16″. A greater accuracy of 1/16″ is not usually required during design or construction, but if you do need a greater accuracy, **Units** is the tab where this can be updated.

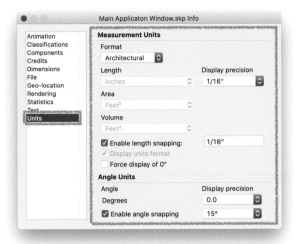

Figure 1.21   Model Info window shown with the Units option activated.

Modeling Tip

The **Decimal** format selection is the only selection where the S.I. units system can be chosen besides the U.S. customary system.

Toolbar and Main Modeling Window

The toolbar and the main modeling window are grouped under the same heading because they are interconnected by space, or lack of it. Every SketchUp user will have to find the right balance of working area versus individual tools located in the toolbar. By adding more tools in the toolbar you have less working space, and you will find out on more than one occasion that every pixel counts. Another issue of having too many tools in the toolbar is the shear clutter that can lead to confusion and a decrease in your productivity. The opposite is true as well; by having less tools in your toolbar your working space will increase, but your productivity will decrease since you will have to go back and forth adding and removing tools. So the question is, What is the right balance?

Figure 1.22 shows an example of balance between tools and working space. This balance satisfies a need for work space based on viewing screen size and the number of tools that are in use on daily basis. The balance in the example was achieved from work experience and separating core tools into two baskets: (1) tools that are used frequently on daily basis and (2) tools that are for modeling work but are used for specific tasks. The ability to understand what tools you use the most can help you further in positioning these tools on

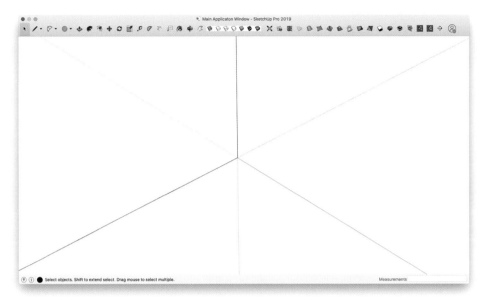

Figure 1.22   The main working window and the balance between tools and usable space.

the toolbar and therefore increase your productivity without sacrificing working area. This is discussed further in the section titled **Tools** later in this chapter.

Measurements Input Box

The **Measurements Input Box** is located at the bottom-right corner of the main modeling window (**Fig. 1.23**). The principal characteristic of the **Measurements Input Box** is that it

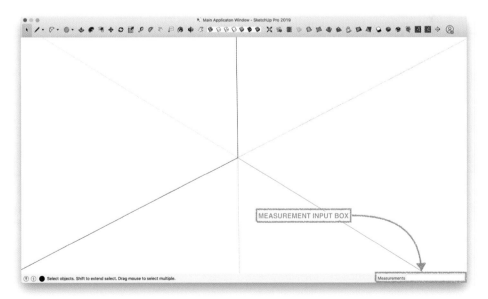

Figure 1.23   Examples of changes in the Measurement Input Box based on tools and required input parameters.

acts as a centralized input location for the entire SketchUp application. Each time a different tool is selected for use, the **Measurements Input Box** changes based on the input parameters required for that tool to function; see **Fig. 1.23** for different input box parameters.

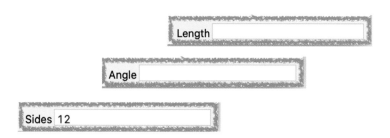

Tools and Toolsets

SketchUp has a wide variety of tools and toolsets available for use by you as the modeler. The following section focuses on the review of the most commonly used tools and toolsets throughout the course of the book. The idea is to set a level plane of understanding and knowledge for the novice and experienced users alike for the most commonly used tools and toolsets. The level plane will help everyone reading this book to effortlessly follow the workflow and the examples in this book.

Before the review, it will be beneficial to go over the procedure for searching and adding tools to the toolbar. Tools can be accessed and added either from the menu bar or directly from the toolbar. To add tools from the toolbar, right-click over the toolbar and from the context menu select the option **Customize Toolbar… (Fig. 1.24)**.

After the selection is made, the tool addition window will appear, which will show a list of all the available tools that come standard with SketchUp (**Fig. 1.24**).

To add tools from the menu bar, click on **View** and select the **Customize Toolbar…** option from the menu (**Fig. 1.25**). Similar to the previous procedure, after the selection is made, the tool addition window will appear, which will show a list of all of the available tools that come standard with SketchUp (**Fig. 1.25**). After this, the addition of the tools is very simple for both procedures, just click and drag the desired tool to the toolbar.

Modeling Tips

(1) Besides the standard SketchUp tools, the tool addition window also holds all the tools from the extensions that are installed on your SketchUp. (2) The tool addition window also allows you to specify how SketchUp will show your tool icons in the toolbar (**Fig. 1.26**). In order to save on work area, choose the **Icon Only** option with **Use Small Size** enabled—this is a personal preference. Besides the **Icon Only**, SketchUp also gives you the **Icon and Text** and **Text Only** options. Try these options and see what works best for you.

Figure 1.24   Adding tools from the toolbar.

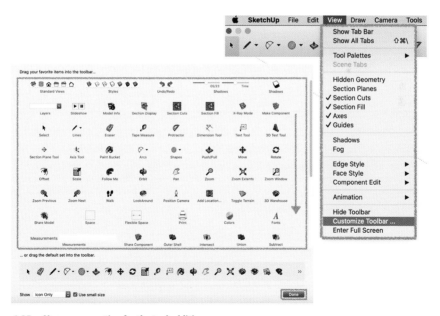

Figure 1.25   Menu access option for the tool addition menu.

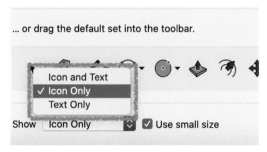

Figure 1.26   View options for tool.

Based on the general function they perform, you can use three major categories to differentiate the tools and toolsets used in this book:

1. **Creation Tools**
2. **Modification Tools**
3. **Assistant Tools**

The **Creation tools** can be further subdivided into two separate categories: **Edge Creators** and **Face Creators**. In the following section, each of these categories of tools and their individual properties are reviewed.

Edge Creator Tools

Under the **Edge Creator** category, we will find the following tools and toolsets: **Line**, **Freehand**, **Arch**, **2 Point Arch**, **3 Point Arch**, and **Pie** (**Fig. 1.27**).

Figure 1.27   Icon-type representation of the **Edge Creator Tools.**

The main purpose of these tools is to generate the most basic element in SketchUp, and that is an edge. Although these tools have the ability to generate surfaces by connecting a minimum of three coplanar edges, they are mostly used to improve or enhance an already-created surface or to give a path of extrusion when three-dimensional shapes are being created.

Line and Freehand Tools

The **Line** and **Freehand** tools are one of the more basic tools in the SketchUp arsenal.

The procedure for creating an edge (line) is very straightforward. The first step is to select the **Line** tool from the toolbar and then to select a starting point by clicking on your left mouse button. The next step is to choose an end point. This can be done in two ways, either by moving the mouse to the desired destination and clicking the left mouse button or entering a distance value in the **Measurements Input Box** (**Fig. 1.28**).

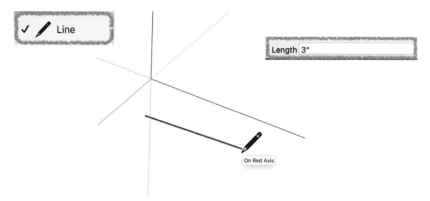

Figure 1.28   Procedure for creating a **Line**.

In order for an edge (line) to maintain the correct axis, SketchUp has built an axis snap by utilizing the arrow keys on your keyboard, which are as follows (**Fig. 1.29**):

1. **Upper Arrow key**: The edge (line) will snap and follow the **Blue** axis.
2. **Left Arrow key**: The edge (line) will snap and follow the **Green** axis.
3. **Right Arrow key**: The edge (line) will snap and follow the **Red** axis.

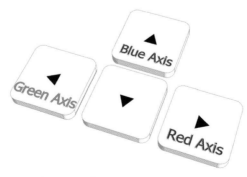

Figure 1.29   **Keyboard buttons and corresponding snap.**

In order to activate the snap (follow) option, prior to selecting an end point, press on the desired arrow key, the edge (line) will appear as a bold line with the corresponding color of the axis (**Fig. 1.30**).

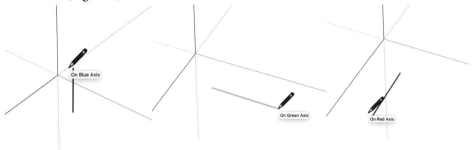

Figure 1.30   Different snap options and the change in color based on the axis of snap.

Similar to the **Line** tool, the **Freehand** tool operates on the same basis. After selecting the **Freehand** tool from the toolbar, press and hold the left mouse button and move it in the desired direction. This will create a freehand edge (line) on your drawings area. Any location where a freehand edge (line) meets in a coplanar way, a face will be created (**Fig. 1.31**).

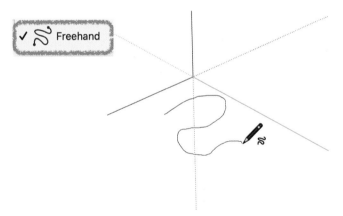

Figure 1.31   Procedure for creating a **Freehand Line**.

The **Arc, 2 Point Arc, 3 Point Arc**, and **Pie** tools are very similar in nature, in that they create a geometric arc. The only differences between these tools are in the number of reference points needed to create the arc and the type of arc they create (circular, semicircular, vs. parabolic—which is the most commonly used arc in the bridge design).

1. **Arc** tool: It creates a circular or semicircular arc. The first step after selecting the **Arc** tool is to specify the center point of the arc, followed by the end point (this will be your radius since it is a circular type of arc). This operation can be completed by pressing the left mouse button. The end point can be chosen either by moving the mouse cursor to the desired distance and pressing the left mouse button again or by entering it manually in the **Measurements Input Box**. The next step is to choose the angle of the arc and create your arc length by either moving the mouse cursor

in the desired rotation and pressing the left mouse button or by entering an angle in degrees manually in the **Measurements Input Box** after selecting the correct vector of movement (**Fig. 1.32**).

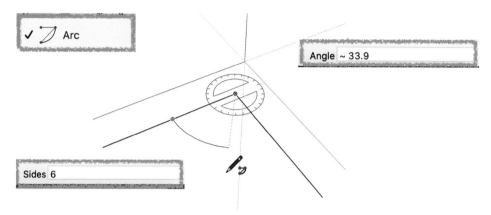

Figure 1.32   Procedure for creating an Arc.

2. **2 Point Arc** tool: It creates a parabolic type of arc from two fixed points and one variable point. The first step after selecting the **2 Point Arc** tool is to specify the start and end point of your arc. This operation can be completed in two ways: by either selecting the start and end points with a left mouse button click or by selecting a start point and typing a length value for the end point in the **Measurements Input Box**. When the start and end points are selected, SketchUp will ask you for the last point, which is the location of the vertex point or bulge as it is specified in SketchUp. The vertex point dimension can be added either by moving the mouse cursor to the desired location or by typing the distance in the **Measurements Input Box** after selecting the correct vector of movement (**Fig. 1.33**).

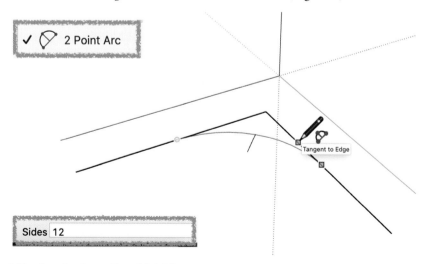

Figure 1.33   Procedure for creating a **2 Point Arc**.

3. **3 Point Arc** tool: It creates a circular or semicircular type of arc from two fixed points and one variable point. The first step after selecting the **3 Point Arc** tool is to specify the start point and a secondary fixed point (the diameter of the circular arc) from which the internal angle will be modified as the mouse cursor is moved. This operation can be completed in two ways: by either selecting the start and fixed points with a left mouse button click or by selecting a start point with a left mouse button click and typing the fixed point distance in the **Measurements Input Box** after selecting the correct vector of movement. When the start and fixed points are selected, SketchUp will ask you for the last value, and that is the internal angle of your arc. The internal angle can be added either by moving the mouse cursor and left-clicking on the desired location or by typing the angle in degrees in the **Measurements Input Box** after selecting the correct vector of movement (**Fig. 1.34**).

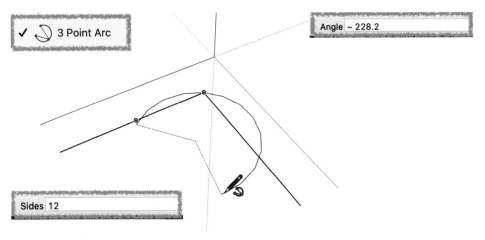

Figure 1.34   Procedure for creating a 3 Point Arc.

Modeling Tip

The **2 Point Arc** and the **3 Point Arc** tools are a big help when setting up a bridge baseline, curvature on a parapet wall, and chamfer on concrete or steel structures.

4. **Pie** tool: It creates a circular or semicircular arc. The first step after selecting the **Pie** tool is to specify the center point of the arc, followed by the end point (this will be your radius since it is a circular type of arc). This operation can be completed by pressing the left mouse button. The end point can be chosen either by moving the mouse cursor to the desired angle or by entering it manually in the **Measurements Input Box**. When the operation is completed, the **Pie** tool will create an arc segment with a face (**Fig. 1.35**).

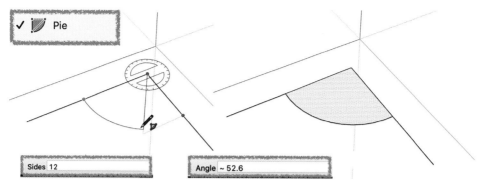

Figure 1.35   Procedure for creating a Pie.

When the **Arc, 2 Point Arc, 3 Point Arc**, and **Pie** tools are selected, SketchUp gives you an option to increase or decrease the number of segments that will be used to create the arc. In order to increase or decrease the number of segments, you can either use the **Measurements Input Box** and type in the new number of segments or press **Option +** to increase and **Option** — to decrease the number of segments. If you use the second option, keep in mind that you add or subtract one segment every time you press the key composition. Also, unlike option 1, option 2 works even when you are in the middle of drawing the arc.

Face Creator Tools

Under the **Face Creator** category, you will find the following tools and toolsets: **Rectangle, Rotated Rectangle, Circle**, and **Polygon**. The main purpose of these tools is to generate faces in the model, which can be further modified in order to produce three-dimensional shapes.

Rectangle and Rotated Rectangle Tools

The **Rectangle** tool is by far one of the most used tools in the SketchUp arsenal of tools for the creation of surfaces, without the worry of your edges being coplanar in nature. The procedure for creating a rectangular surface is as follows (**Fig. 1.36**):

1. Select the starting point of your rectangle by clicking on your left mouse button.
2. Moving diagonally with the mouse, select the end point of the rectangle by clicking the left mouse button—note the change in distance values at the **Measurements Input Box** while moving the mouse to the end point.

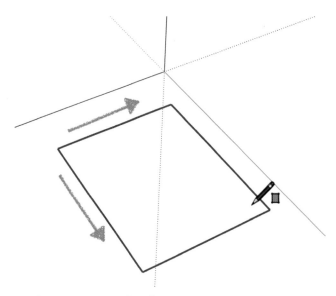

Figure 1.36   Procedure for creating a **Rectangle** surface.

3. Accurate distance values can be manually input by typing them in the **Measurements Input Box** with the following input convention 0(Units), 0(Units) (**Fig. 1.37**).

Dimensions 1/4", 5/16"

Figure 1.37   Example of input convention for a rectangle.

In 2015, SketchUp introduced a new tool called **Rotated Rectangle**. The tool allows you to create a rectangle on an angle or to start a rectangle from a predetermined edge: basically it allows for more control on how you create your surface. The procedure for using the **Rotated Rectangle** tool is as follows:

1. Select the starting point of your rectangle by clicking on your left mouse button. SketchUp will ask for a secondary point. The secondary point has two main purposes: it establishes the length of a side of the rectangle and establishes an axis about which the rectangle is rotated (**Fig. 1.38**).

2. When the start point and the rotational axis are selected, move the mouse in order to select the end point and the angle at which the rectangle will be created—note the change in distance values at the **Measurements Input Box** while moving the mouse to the end point.

3. Accurate distance values and rotational degrees can be manually input by entering them in the **Measurements Input Box** with the following input convention 0(Distance Units), 0(Rotational Degrees).

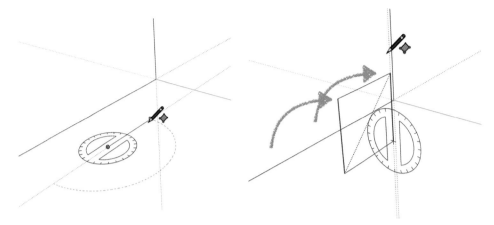

Figure 1.38   **Procedure for creating a Rotated Rectangle.**

Modeling Tip

Unlike the **Rectangle** input convention where you choose the length and width of the rectangle, in the **Rotated Rectangle** the choice is only for one side of the rectangle and the degree at which the rectangle will be rotated. The secondary point that was chosen under step 1 sets the initial length of the rectangle.

Circle and Polygon Tool

As was mentioned earlier in this chapter, SketchUp is a three-dimensional drafting application that is based on edges and surface. Because of this basic concept of operation, a circle in SketchUp is created from multiple straight edges that are interconnected with each other in order to make a circular surface (**Fig. 1.39**). In reality, SketchUp produces a circular "polygon" rather than a "real" circle.

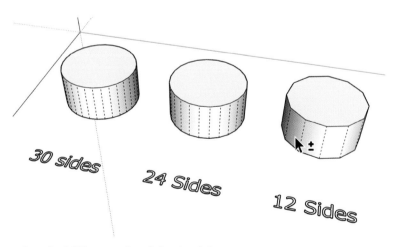

Figure 1.39   **Example of different number of edges in a circle.**

The procedure for using the **Circle** tool is as follows:

1. Select the **Circle** tool from the toolbar and bring your attention to the **Measurements Input Box** (**Fig. 1.40**). In SketchUp the default number of edges in a circle is set to 24. What this number means is that the circle will be created from 24 straight lines, interconnected in a circular shape.

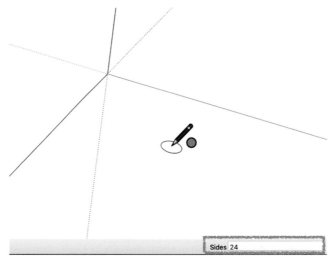

Figure 1.40   SketchUp request for number of sides.

2. Select the center point from which the circle will originate, by clicking on the left mouse button. Select the end point of the circle by moving the cursor away from the center point in any desired direction and again clicking the left mouse button—note the change in radius distance values in the **Measurements Input Box** while moving the cursor to the end point (**Fig. 1.41**).

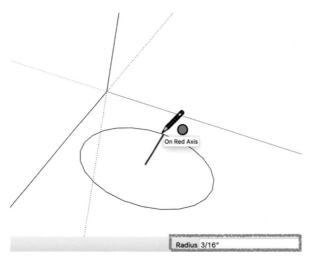

Figure 1.41   Adding radius to the circle in the Measurements Input Box.

3. An accurate radius value can be input manually by entering it into the **Measurements Input Box** with the following input convention 0(Radius Distance Units) (**Fig. 1.41**).

Modeling Tip

The default edge count of 24 edges can be changed by inputting a new number in the **Measurements Input Box** prior to selecting the center point of the circle. Increasing the number of edges will also increase the demand on your computer and slow down the response time of SketchUp, especially if you are dealing with multiple circular shapes, as it would be the case for foundation piles at bridges, light or electrical poles, fences, and so on.

The **Polygon** tool: The procedure for using the **Polygon** tool is as follows:

1. Select the **Polygon** tool from the toolbar and bring your attention to the **Measurements Input Box** (**Fig. 1.42**). In SketchUp, the default number of sides (edges) for the **Polygon** tool is set to 6 (hexagon).

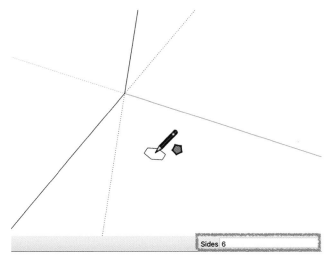

Figure 1.42   SketchUp request for number of sides during the initial stages of a polygon creation.

2. In order to change the number of sides, select the **Polygon** tool from the toolbar and enter the new desired number of sides in the **Measurements Input Box**.

3. Select the center point of your polygon by left-clicking on the drawing area. In order to select the end point or the desired radius, move the mouse cursor away from the center point—note the change in radius distance values in the **Measurements Input Box** while moving the cursor to the end point (**Fig. 1.43**).

4. An accurate radius value can be input manually by entering it into the Measurements Input Box with the following input convention 0(Radius Distance Units).

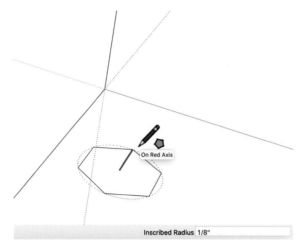

Figure 1.43 **Procedure for creating a Polygon.**

Modeling Tip

The **Polygon** tool in SketchUp creates circumscribed polygons; basically the polygon is inscribed inside of an imaginary circle that passes through all the vertices of the polygon. This is very important from an accuracy standpoint, since the radius (circumradius) is measured to the tip of the vertex and not to the edge (**Fig. 1.44**).

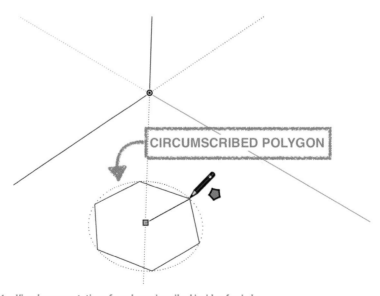

Figure 1.44 **Visual representation of a polygon inscribed inside of a circle.**

Modification Tools

Under the **Modification** category, you will find the following tools and toolsets: **Push/Pull**, **Follow Me**, **Offset**, **Move**, **Rotate**, and **Scale** (**Fig. 1.45**).

Figure 1.45   Example of Modification Tools.

The main purpose of the **Push/Pull** and **Follow Me** tools is to generate three-dimensional shapes from surfaces. The procedure to use the **Push/Pull** tool is very simple: after you complete the selection process for the tool from the toolbar, you can proceed to select, by left mouse click, any valid surface that you want to extrude in three dimensions (**Fig. 1.46**). The amount of extrusion can be either estimated by moving the mouse cursor away from the surface that you are extruding from or by specifying it by entering an exact distance in the **Measurements Input Box** (**Fig. 1.46**).

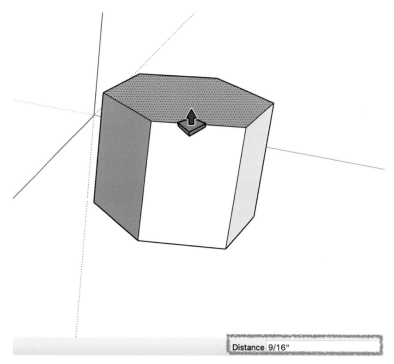

Distance 9/16"

Figure 1.46   **Procedure for creating extruding a surface using the Push/Pull tool.**

The **Push/Pull** tool will only extrude a three-dimensional shape that is perpendicular to the starting surface.

The **Follow Me** tool extrudes three-dimensional surfaces by following a path of extrusion. In order to use the **Follow Me** tool you have to satisfy two requirements: a valid surface and a path. The procedure to use the **Follow Me** tool is very simple: after the tool is selected from the toolbar, you can proceed to select the valid surface by left-clicking on it and the path you would like to extrude the surface by (**Fig. 1.47**).

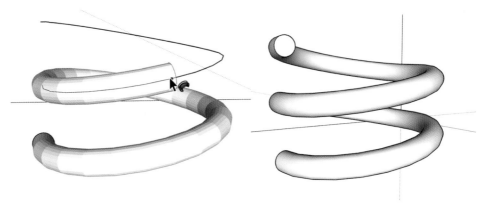

Figure 1.47   Example of using the **Follow Me** tool in order to create a spiral reinforcement bar.

Unlike other tools in SketchUp, the **Follow Me** tool does not have an option to enter an exact distance of extrusion. The initial distance that you assign to the path becomes the final extruded length of the three-dimensional object. Imagine the path as the skeleton of your extrusion, which also limits the length.

Some examples of uses for the **Follow Me** tool are when generating complex three-dimensional models of post-tensioning cable, reinforcement bars, pipe elbows with different degrees of bend, ornamental details, and so on. See **Chapter 7—Modeling of Various Bridge Components and Accessories** for more an in-depth discussion on the creation of these objects.

The **Offset** tool functions only on surfaces and will not offset edges by themselves. The main benefit of this tool is that you can offset a series of edges on a surface, internally or externally, in order to produce new surface(s) for addition or subtraction purposes (**Fig. 1.48**).

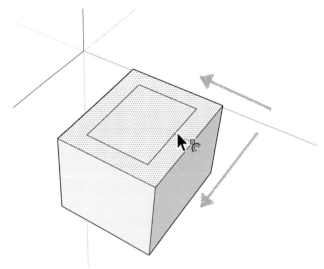

Figure 1.48   Example of an offset on top of a surface.

The **Move** and **Rotate** tools have two primary functions. The first function is to move and/or rotate edges, surfaces, or entire three-dimensional objects around the working area (**Fig. 1.49**).

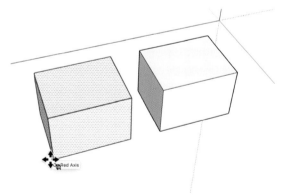

Figure 1.49   Example of an object move using the Move tool.

The second function is to act as object copiers. In order to copy an object, which can be an edge, surface, or collection of surfaces, use the following steps:

1. Select the object you want to make a copy of.
2. Click on the **Move** or **Rotate** tool and then by holding down the **Option** key, left-click anywhere on the object or outside the object as a reference point.

3. Release the **Option** tool and with your mouse, drag or rotate the copy of the object to the new position (**Fig. 1.50**).

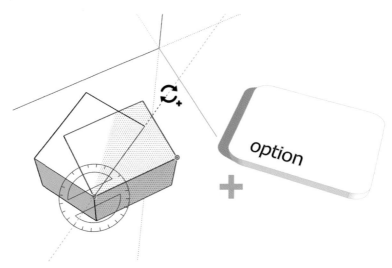

Figure 1.50   Procedure for creating a copy using the Rotate tool.

Modeling Tip

In order to snap to an axis during your move, rotate or copy work, use the upper arrow key to snap to the Blue axis, left arrow key to snap to the Green axis, and the right arrow key to snap to the Red axis. The bottom arrow key allows you to snap to a predetermined surface or edge (**Fig. 1.51**).

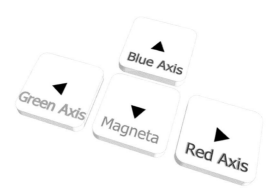

Figure 1.51   Arrow keyboard keys and the associate snap option.

The **Scale** tool has two primary functions. The first function is to scale a surface up or down or a collection of surfaces (an object). The necessary steps to scale a surface or an object are as follows:

1. Select the object you would like to scale.
2. Click on the **Scale** tool, which will automatically create a set of scalable points (green squares) around your object (**Fig. 1.52**).

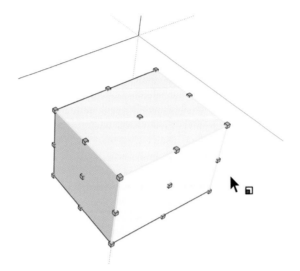

Figure 1.52   Visual representation of scalable point when the **Scale** tool is activated.

3. Depending which scalable point (green square) you left-click and drag will also determine if the scale is uniform in nature or along a specific axis. In order to have a uniform scale, select and hold the corner point. Selecting any other point, besides the corner points, will produce scaling based on a specific axis only (**Fig. 1.53**).

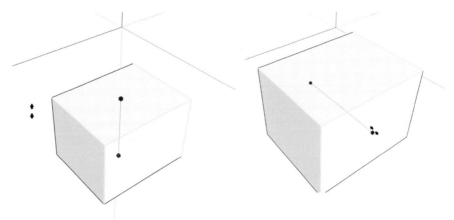

Figure 1.53   Example of scaling based on a specific axis.

4. Scaling can be done manually by moving the mouse toward or away from the object or by entering a specific scaling fraction in the **Measurements Input Box**.

The second function of the **Scale** tool is to act as a mirror tool. SketchUp does not have a designated tool for mirroring object, and therefore, the **Scale** tool also performs that specific task. The steps to mirror an object are the same as specified earlier, with one difference. The object will be mirrored and not scaled when the value in the **Measurements Input Box** reads −1.0 (**Fig. 1.54**).

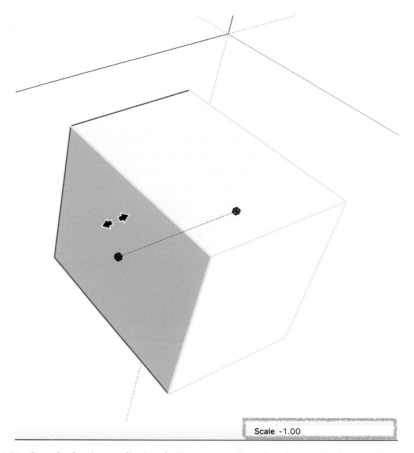

Scale -1.00

Figure 1.54 Example of a mirror application: the **Measurement Input Box** shows −1.0 value which is an indication that the object was mirrored.

Assistant Tools

Under the **Assistant** category, you will find the following tools and toolsets: **Styles** toolset, **Tape Measure, Protractor, Dimension, Text, Pan, Orbit, Zoom, Sandbox** toolset, **Shadows, Add Location…, Toggle Terrain, Section** toolset, Fog and Scenes.

The main purpose of the tools in the **Assistant** category is to provide a visual aid to the user during the three-dimensional modeling work. The **Sandbox** toolset, **Add Location…**, and **Toggle Terrain** are reviewed in **Chapter 9—Site Modeling and Use of Site Creation Tools**, utilizing an example.

The **Styles** toolset offers a set of visual aid to the user during the three-dimensional modeling work **(Fig. 1.55)**.

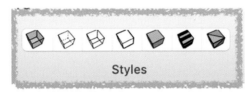

Figure 1.55   **Styles** toolset: each visual style shows a different aspect of your model.

There are a total of seven visual aids, and they are as follows:

1. **X-Ray**: All hidden lines and surfaces are shown for the three-dimensional object.

2. **Back Edges**: Hidden lines are shown as dashed lines for the three-dimensional object.

3. **Wireframe**: Surfaces are not shown for the three-dimensional object.

4. **Hidden Lines**: Hidden lies and textures are not shown for the three-dimensional object and all the surfaces are colored white.

5. **Shaded**: Only textures will not be shown on the three-dimensional object.

6. **Shaded with Texture**: Represents how the three-dimensional objects are visually shown in SketchUp.

7. **Monochrome**: All color or texture will not be shown on the three-dimensional object.

Tape Measure and Protractor

The **Tape Measure** and the **Protractor** tool provide two main functions: to create assist lines and to measure distances on the model **(Fig. 1.56)**. Assist lines are very helpful when working with any type of geometry (complex or noncomplex) because they reinforce the modeling process as placeholders, offset position locator, path locator, and so on. Furthermore, assist lines do not get entangled with the geometry as would be the case for a regular edge created by the **Line** tool.

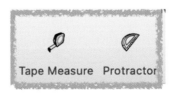

Figure 1.56   Visual representation of the **Tape Measure** and **Protractor** tool.

The **Tape Measure** tool can only create perpendicular lines assist based on the offset edge of an object or the axis (**Fig. 1.57**). In addition, the **Tape Measure** can also place a guide point based on an object point. The offset location of the assist line can be entered manually in the **Measurements Input Box** or achieved by visual guidance.

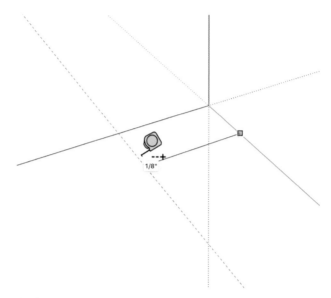

Figure 1.57   Example of assist line placement with the **Tape Measure** tool.

The **Protractor** tool can create an assist line on any angle by selecting a base path and rotating that base path to the desired angle (**Fig. 1.58**). Utilize the keyboard arrows to snap to an axis (upper arrow key to snap to the Blue axis, left arrow key to snap to the Green axis, the right arrow key to snap to the Red axis, and the bottom arrow key to snap based on the surface location). The angular rotation can be entered manually in the **Measurements Input Box** or achieved by visual guidance.

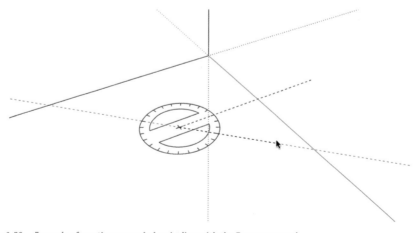

Figure 1.58   Example of creating an angled assist line with the **Protractor** tool.

Dimension and Text

The **Dimension** and **Text** tools assist the user to enter dimensional or textual information on the three-dimensional model. Changes to the font properties can be accomplished
in two ways: (1) selecting either the dimension or text by a double left- click and, with a
right-click on the mouse button, selecting **Font** then **Show Fonts** from the context menu or
(2) activating the **Entity Info** window by selecting **Window** then **Entity Info** from the
menu bar and then clicking on the **Font…** button. Either option chosen will activate the
Fonts window (**Fig. 1.59**).

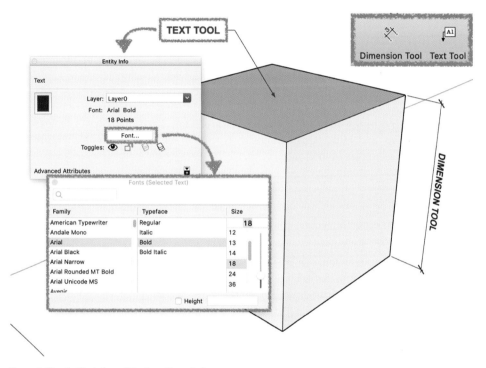

Figure 1.59   **Entity Info** and **Fonts** option windows.

Pan, Orbit, and Zoom

The **Pan, Orbit,** and **Zoom** tools allow the users to orient themselves in the main modeling
window. All three tools can be activated from scroll wheel–type mouse, directly from the
toolbar or the main menu bar.

1. The **Pan** tool allows you to move the view vertically or horizontally. Depending on
    the mouse type, and availability of buttons, the **Pan** tool can be activated by clicking
    on a designated additional mouse button (the mouse I use on a daily basis has two
    additional buttons besides the standard ones). The **Pan** tool can also be activated
    from the menu bar by clicking on **Camera** and selecting the **Pan** option or directly
    from the toolbar by selecting the **Pan** icon.

2. The **Orbit** tool allows you to orbit around a three-dimensional object or on the modeling plane. In order to activate the **Orbit** tool from a scroll wheel–type mouse, click and hold the designated additional mouse button (the mouse I use on a daily basis has two additional buttons besides the standard ones). The **Orbit** tool can also be activated from the menu bar by clicking on **Camera** and selecting the **Orbit** option or directly from the toolbar by selecting the **Orbit** icon.

3. The **Zoom** tool allows you to zoom in or out of the modeling area. The zoom tool is automatically activated when scrolling in and out on a scroll wheel–type mouse. The **Zoom** tool can also be activated from the menu bar by clicking on **Camera** and selecting the **Zoom** option or directly from the toolbar by selecting the **Zoom** icon.

The Inference Engine

SketchUp has a highly intuitive inference engine, which is of great help to modelers, regardless of whether the work is on complex or noncomplex three-dimensional models. Unlike other CAD applications, the inference engine in SketchUp tries to predict your need for inference during your three-dimensional modeling. Each inference is separated by a different color and a text explanation what type of an inference is possible. **Figure 1.60** shows different examples of inference.

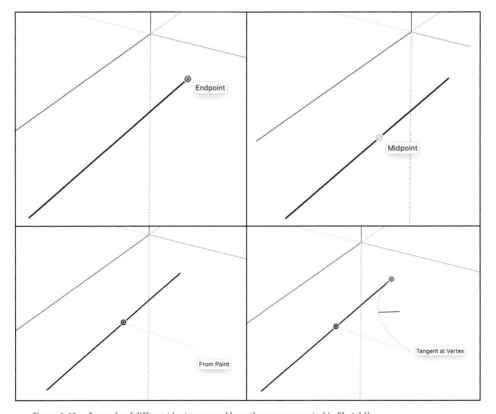

Figure 1.60   Example of different instances and how they are presented in SketchUp.

Components and Groups

This section reviews components and groups, which represent a very important aspect of SketchUp. There are many questions with respect to components and groups: When would be the best to use one versus the other, or what is the overall benefit each one contributes to your three-dimensional model? The intention of this section is to provide some guidance about, as well as clarify any misconceptions regarding, groups and components.

The first questions that need to be clarified are, What is a group, and what is a component? Components and groups share a key fundamental origin: they both represent a collection of geometry; you can visualize them as storage boxes. As you create a three-dimensional object, and you make it into a component or a group, you are technically putting all the parts and pieces of that object (lines, faces, color, etc.) into this imaginary storage box. After SketchUp packs your three-dimensional object into this imaginary storage box, it will request for you to assign a name or a label to your component or group in order to make it unique. You can think about it in terms of writing the content of a storage box so you can identify it later when you need to use it.

The difference between a group and component storage box lies on what each represents in the model. A group represents a single occurrence (think about it as one complete storage box) of geometry. Unlike a group, a component represents a tag of named occurrences of geometry (you no longer have a box but only a placeholder or a tag for the box that was created).

This brings up the next set of questions: When would you use a group or a component, and what is the benefit of a group or component? It is recommended to group objects when you do not intend to make copies of the same group multiple times, with the benefit that it will stop the "stickiness" between objects. The term "*stickiness*" is used to explain the interaction between edges and surfaces in SketchUp. On the other hand, it is recommended to use components when you intend on using the same object multiple times throughout the drawing or on any future drawing and when there is a possibility of multiple revisions. The benefits with components are prevention of accidental modification of the object (stopping the stickiness property), saving the component for future use, easy modification or full replacement of components, and, most important, decreasing the file size of the drawing. Components can always be updated, which is discussed in **Chapter 7—Modeling of Various Bridge Components and Accessories**. These four benefits can be reviewed separately and a parallel can be drawn between a group and a component.

1. Prevention of Accidental Modification

One of the fundamental properties of SketchUp is stickiness or connection of surfaces (this concept was covered in more detail under **Chapter 1, The SketchUp Concept—Connection (Stickiness) of Surfaces**). This property can be highly helpful in some instances, and in others it can hinder your progress. By making a group or a component from an object, you stop this property from occurring. The reason for this is when you have a group or a component you create a separate entity; the best way to think about it is you have created a universe within a universe where some properties will function and others will not. A good example of this stickiness or connection of surfaces property is how you can still draw on the group or component that you created but they will not be blended within them.

2. Easy Modification and Full Replacement Option

Unlike a group, a component can be easily modified or fully replaced for another component. This is particularly useful during preliminary design stages of projects and work planning operations. It also increases SketchUp's response time by adding low-polygon objects which later will be replaced by high-polygon objects. There are two procedures for replacing and/or updating components.

In order to replace a currently used component in a drawing with either a high-polygon count duplicate or entirely replace the component with a different component, first you have to activate the **Components** inspection window. The **Components** inspection window can be activated by selecting **Window** and then selecting the **Components** option from the main SketchUp menu (**Fig. 1.61**).

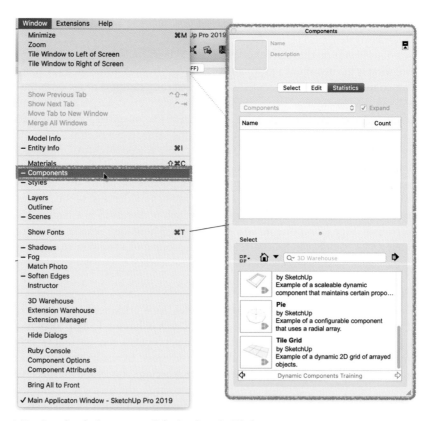

Figure 1.61   Accessing the **Components** dialog box from the **Window** menu.

The next step is to select all the instances from the component that you want to replace. You can accomplish this task in two ways, either by selecting all the instances individually in the drawing (**Fig. 1.62**) or by right-clicking on the component in the **Select** box, which is located under the **Components** inspector window, and selecting **Select Instances** from the context menu (**Fig. 1.62**).

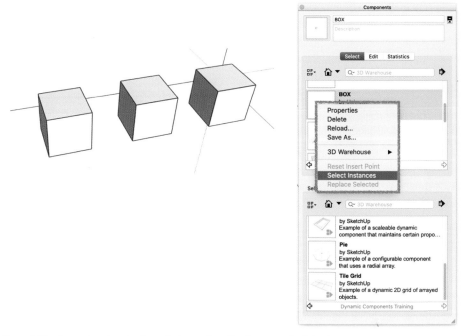

Figure 1.62   Example on how to choose instances from the **Components** dialog box.

After all the instances are selected, hover over the new component that you would like to use, and by right-clicking on it, select the **Replace Selected** from the context menu (**Fig. 1.63**). This will replace all the instances of the component with the new component chosen.

Figure 1.63   Example on how to replace instances from the **Components** dialog box.

The second procedure is mainly used to update a single instance of a component already used in the drawing, or you can also replace the instance with a new component. The procedure for accomplishing this task is to first select the instance that you want to update or replace. When the instance is selected, right-click over it, and from the context menu select **Reload…** (**Fig. 1.64**).

SketchUp will show you an information window, informing you that the component file has not been changed and whether you want to replace the component with a new one. If the component file has been changed, you will not see this window and the instance will be updated from the new component file. Since the component file has not changed, you will select the option **Yes** and select the new component from the options list that you want to use for the selected instance.

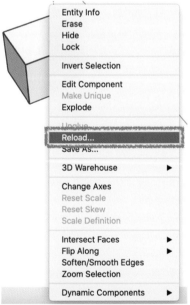

Figure 1.64   Example on how to replace an instance from the context menu.

Modeling Tip

When you update or replace an instance or an entire component, the new component will be imported with all the properties as it was saved (specific scaling that was applied, axis location, possible rotation, etc.).

3. Building a Personal Library

SketchUp has a default library of very basic components that come standard with the application. The default library can be found under the **Components** inspection window (see **2. Easy Modification and Full Replacement Option** on the procedure to gain access to the **Components** inspection window). As you improve your modeling skills and you gain more "real-world" knowledge, you will start to increase your own library of components. This library will be a mixed batch of bridge components, construction equipment components, standard construction detail components, and so on.

The first and very important step is to organize your components by separating them into different folders based on types (**Fig. 1.65**).

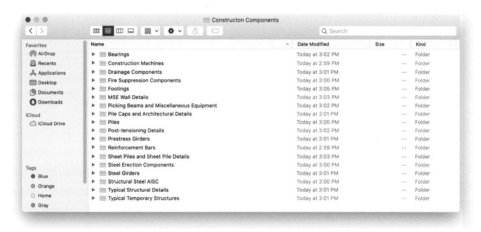

Figure 1.65   Example of a components library. It is very important to properly separate your components for ease of review and access.

This can be a very tedious task if you have not done it before, but it will be very helpful in the end, when you are trying to use your own library. When you are completed with your organization, left-click on the **Details** icon located on the **Component** inspection window and select the **Create a new collection…** option from the context menu (**Fig. 1.66**).

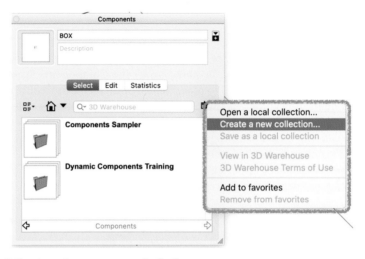

Figure 1.66   Options to create a new component collection.

When the selection window is open, navigate to your collection and click on the **New** button (**Fig. 1.67**). Your component library will be loaded in the **Component** inspection window for your use.

Figure 1.67   Choose the correct library of components to be loaded into SketchUp.

Modeling Tip

At the selection window, activate the **Add to favorites** option. Activating this option will add your library in the favorites.

4. Decreasing the File Size

Components, unlike groups, will decrease your file size and increase your productivity when building a model that requires multiple copies (instances) of an object. The best example of this is a reinforcement bar or a bolt in a girder splice to mention a couple. The reason for this is found in the fundamental difference between components and groups, which was explained thoroughly in the beginning of this section.

When a group is copied multiple times in the model, you are making a copy of the entire storage box (the analogy from the beginning of this section) and will have the complete set of all the properties, geometry, and everything else that is part of that group. Unlike a group, when a component is copied, you are only copying a predetermined instance or a tag (the analogy from the beginning of this section) of the component.

CHAPTER 2

SketchUp and Resources

3D Warehouse   Web-Based Resources
Extension Warehouse and Extensions

SketchUp's greatest strength lies in the large community of its users: people like you, me, and hundreds of other individuals and companies who enjoy working in SketchUp on a daily basis for professional or personal work. This enthusiasm toward SketchUp can be seen in two main SketchUp gateways for information and applications, which are the **3D Warehouse** and the **Extension Warehouse**. This chapter explores the different locations for resources that are readily available for SketchUp users.

3D Warehouse

The **3D Warehouse** gateway represents one of the biggest libraries for SketchUp three-dimensional models. The entire content of models in the **3D Warehouse** library is fully made by contributions from individual users or companies. Anyone can upload or download any model from the **3D Warehouse** directly into their drawings. It should be pointed out that the library, along with its content, is free for all users, and as such the models are not checked for physical accuracy. This might not sound like a problem, but it can become one, especially if you are making site-accurate three-dimensional models. One example where this can become a problem is when creating a model for steel erection purposes. The accuracy of the crane, specifically the accurate position of the pin with regard to the crane body can make a difference in the actual capacity and in turn create a planning problem for the erection of the girders. A rule of thumb is to check who produced the model (company versus individual) and always check it for accuracy.

There are two options to access the **3D Warehouse** library. Option one is to click on the **3D Warehouse** tool located in the toolbar (**Fig. 2.1**). Option two is to access the **3D Warehouse** from the main menu by clicking on **Window** and selecting the **3D Warehouse** option (**Fig. 2.1**).

Figure 2.1   Option on how to access the 3D Warehouse library.

When either option is selected, the **3D Warehouse** window will appear from which you can search for the desired content or specific company (**Fig. 2.2**). Also this window allows you to upload any content that you would like to by selecting the upload content icon (**Fig. 2.2**).

Figure 2.2   3D Warehouse start window.

Extension Warehouse and Extensions

The **Extension Warehouse** represents one of the largest digital distribution platform or library for SketchUp extensions. The library is developed and maintained by SketchUp for the exclusive use by the modelers working in the application. Besides SketchUp, the majority of the extensions are developed and published by individuals and other companies in order to increase the productivity and to ease the three-dimensional modeling process. Unlike the **3D Warehouse**, the **Extension Warehouse** is a mixture of free and for-purchase extensions.

There are two options to access the **Extension Warehouse** library. Option one is to click on the **Extension Warehouse** tool located in the toolbar (**Fig. 2.3**). Option two is to access the library from the main menu by clicking on **Window** and selecting the **Extension Warehouse** option (**Fig. 2.3**).

Figure 2.3   Different options on how to access the Extension Warehouse.

Regardless of which access option you choose, the **Extension Warehouse** window will appear (**Fig. 2.4**), from which you can search for extensions based on the desire category or by specific name. The installation procedure is very simple and self-explanatory.

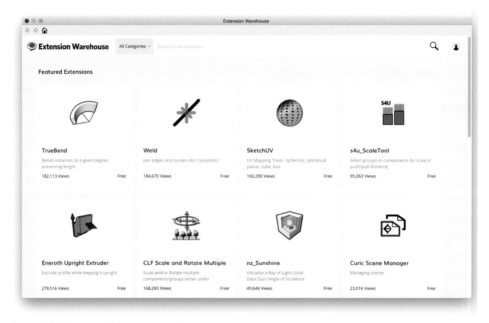

Figure 2.4   Extension Warehouse start window.

If you feel that you can extend the capabilities of SketchUp, and coding is somewhat of a hobby, you too can contribute to the **Extension Warehouse**. There are several very strict steps that have to be taken in order for anyone to upload an extension to the library. The reason for this is overall safety to the design community. Any submittal will go through thorough checks by SketchUp for viruses, code compilation, and so on. More on how to upload an extension can be found on the SketchUp help website.

Web-Based Resources

Besides the **3D Warehouse** and the **Extension Warehouse**, which represents resources made available by the developer of the application, there are also many other, equally good, third-party options on the World Wide Web for SketchUp resources. In closure of this chapter, the following are some of my favorite third-party options for resources and where they can be found on the World Wide Web:

- **SketchUcation** (https://sketchucation.com) is a large community of SketchUp users. **SketchUcation** has a variety of resources, from extensions to models, textures, books, and a large forum where you can gain additional experience from more advanced users.

- **SketchUp Community** (https://forums.sketchup.com) and **SketchUp Help** (https://help.sketchup.com/en) are two great forums for question and help on a large variety of SketchUp topics.

In closing, the following are a few very important points to keep in mind:

- There are many other resource options on the World Wide Web. Explore and have fun; you never know what you can find to make your life easier.

- Use common sense during extensions installation. Never install extensions from sites that you are not one-hundred percent sure they are legitimate.

- Think about SketchUp clutter, too many installed extensions can impact the amount of free drawing space.

- Have fun with the extensions and models, and try to make some yourself; share what you know; the SketchUp community becomes stronger if everyone participates and shares the knowledge you have!

Importing Files, AutoCAD, and SketchUp

This chapter is dedicated to one of the most important options for any modeling or drafting program, and that is the ability to import different types of files. The ability to import files is the ability to save time, and time in the construction industry is everything. Time allows you to maintain a steady flow of submittals in order to keep up with the construction schedule and budget, which will allow you to complete the construction project.

Any project in the heavy civil industry relies on multiple parties to be successful. From the designer to the suppliers and installers of structural steel, steel reinforcement, concrete, drainage, and fire suppression systems—everyone plays a vital cog in the grand engine which moves the project to completion. As construction engineers and members of this driving engine, our sole responsibility is to filter all these parts by combining the shop drawings with the construction drawings and making sure that they fit correctly at the right moment in the project.

Different suppliers and installers use different software systems in order to design and manufacture their portion of the work on a project. Commonly known software packages like Autodesk (AutoCAD, Revit, InfraWorks, 3ds Max, etc.), Bentley (Microstation), Tekla, and many other custom-type software packages are being used for this purpose on a daily basis. The good news is that SketchUp is able to import a wide variety of file extension types, which can be very helpful during the modeling process. The ability to take a file, accurately import it into SketchUp, create a three-dimensional model from that file, run a clash analysis, and use it in the field during construction can be a game changer for small and large construction projects alike.

The following several sections will dive into the types of files that are available for import into SketchUp, the interface between AutoCAD and SketchUp, how to properly prepare an AutoCAD file, and a step-by-step workflow on how to properly import it and use it in your model. The discussions and examples focus primarily on AutoCAD files since they are the most commonly used in the heavy civil industry.

Importing Files

SketchUp gives you the ability to import a wide variety of file types into your three-dimensional model (**Fig. 3.1**).

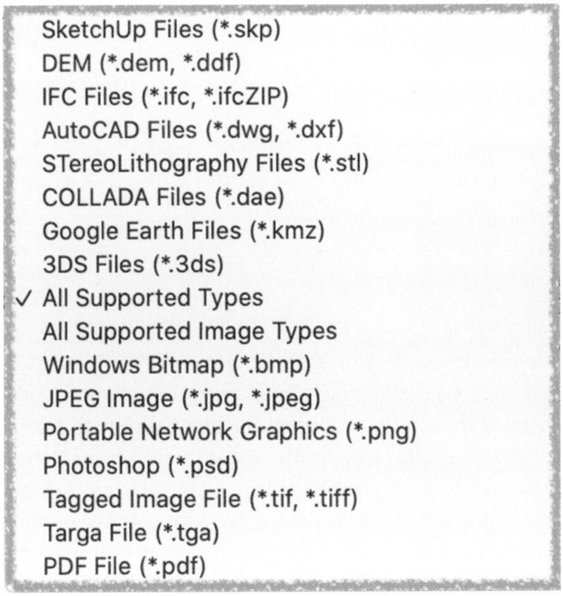

SketchUp Files (*.skp)
DEM (*.dem, *.ddf)
IFC Files (*.ifc, *.ifcZIP)
AutoCAD Files (*.dwg, *.dxf)
STereoLithography Files (*.stl)
COLLADA Files (*.dae)
Google Earth Files (*.kmz)
3DS Files (*.3ds)
✓ All Supported Types
All Supported Image Types
Windows Bitmap (*.bmp)
JPEG Image (*.jpg, *.jpeg)
Portable Network Graphics (*.png)
Photoshop (*.psd)
Tagged Image File (*.tif, *.tiff)
Targa File (*.tga)
PDF File (*.pdf)

Figure 3.1   List of supported files for import by SketchUp.

You can separate these available file types into three separate groups or placeholders:

1. **Group 1** represents a collection of standard image files: ***.bmp, *.jpg, *.jpg, *.jpeg, *.png, *.psf, *.tif, *.tiff,** and ***.tga.** The ability to import these files can be very useful when you want to import location pictures, material patterns, and so on.

2. **Group 2** represents a collection of core import files that are needed to accurately and timely create three-dimensional models by importing required information. These core import files are as follows:

   - ***.dem, *.ddf** (digital elevation model particularly important for terrain import): these file options will be discussed more in **Chapter 9—Site Modeling and Use of Site Creation Tools.**

- ***.ifc, *.ifcZIP** (BIM-type files).
- ***.dae** (digital asset exchange files used to transfer data between three-dimensional software; SketchUp, Maya, 3ds Max, and Rhino).
- ***.kmz** (Google Earth file, particularly important when importing locations): we talk more about the file type and options in Chapter 9-Site Modeling and Use of Site Creation Tools.
- ***.3ds** (Autodesk 3ds Max files).
- ***.dwg, *.dxf** (AutoCAD files, important for the ability to import construction and shop drawings into your three-dimensional model): this is discussed more in depth in the following sections.
- ***.pdf** (Portable Document File): although ***.pdf** files are not, per se, drafting-type files, they are included in **Group 2** since most of the drawings can be found in this particular format.

3. **Group 3** represents files that are used for stereolithography, also known as 3D printing. SketchUp is able to import ***.stl** file types that can be highly useful when working on construction projects with highly complicated post-tensioning tendon locations and when there is the need to physically see them in three dimensions.

Overview—AutoCAD Files and SketchUp

Before you can review the workflow and the accompanying example for importing AutoCAD files, it is very important to give guidance and expand on the steps that are necessary to be taken in order to prepare AutoCAD files for importing into SketchUp. Another equally important topic that is discussed in this chapter is the most commonly found issues during import operations and how the interface between AutoCAD and SketchUp works.

Preparing Your AutoCAD File for Importing

The following is a simple list on how to get started and prepare an AutoCAD file for import into SketchUp. Most of the points in the list are from personal experience and are there to import the drawing correctly, reduce the importing file size, and therefore minimize the starting size of the SketchUp file. This is particularly important when preparing to work with very large base construction AutoCAD files that besides the project also contain a large variety of survey information, local infrastructure information, site elevation contours, horticultural information, and so on. Keep in mind that the import file should only be a base, and therefore, your SketchUp file will only grow in size when you start three-dimensional modeling. By no means is this a definitive list; it is encouraged for everyone to tailor this list to your own needs and even expand on it as you wish from your own experience.

1. **Make a Copy of the AutoCAD File**: The AutoCAD file will be heavily modified following the steps outlined in this chapter. You should not use the original

AutoCAD file that you plan to import into SketchUp, but make a copy that you can modify without the worry that you might lose your original drawing.

2. **Delete all Noting, Dimension, and Hatch Objects**: Delete all dimensions, leaders, text, tables, and hatches—basically everything besides the drawing you want to use for modeling purposes. This is an important step because SketchUp does not recognize noting, dimensions, and hatch objects; therefore, they won't be imported into your drawing, but it will take system resources in order to read them and remove them. Besides, your import file size will be decreased substantially by removing these objects.

3. **Delete all Unnecessary Blocks**: Delete all unnecessary blocks that are present in your AutoCAD drawing, for example, trees and other horticultural blocks, transportation blocks, traffic signs, and the like. These will create a lot of lines when exploded and will drastically slow down your computer during your three-dimensional modeling work. Besides, most of the previously mentioned blocks will be re-created in three dimensions, so there is no need for them to be present in your AutoCAD file.

4. **Consolidate the Drawings**: Use the "Zoom-All" function in AutoCAD to check if multiple revisions or other drawings are present in model-space. This is a good practice, especially if you are working with drawings that were received from companies that you collaborate with on a project. This is also true for intercompany files since your coworkers, who may be other engineers or drafters, tend to "hoard" details on the drawing. The issue with having details or other drawings on the same model-space is that when the AutoCAD file is imported to SketchUp, since AutoCAD's model-space is an endless plane, you can have drawings spaced at large distances. When the file is imported into SketchUp, SketchUp will group everything together and place the grouped file at the origin. By doing so, especially if you are not aware of other drawings, it will be very hard to find the drawing that you want to import. Another issue with having other drawings besides what you want to import is file size considerations that are mentioned under points 2, 3, and 5.

5. **Consolidate and Delete Layers**: Although layers are transferable between AutoCAD and SketchUp, they are not needed for what the AutoCAD file is intended for, as a base drawing—this is discussed in more detail in **Part 1—Importing of AutoCAD Files** exercise. A large number of layers present in an imported drawing will also increase the overall file size of your SketchUp file before you even start your three-dimensional modeling. You should move all the content to one layer and delete or purge the rest of the layers from AutoCAD prior to importing the file into SketchUp.

6. **Explode the AutoCAD Drawing**: Explode the drawing to simple lines and arcs. There are times that SketchUp will not properly read all the content in the importing files, which is because of objects that were imported previously from other software and "stitched" into the AutoCAD drawing, objects made into groups, blocks, or the drawing that contains a lot of polylines or splines. This is not always the case, but it can occur when importing files.

7. **Correct Units**: This last point is to check and understand what units the importing file is drawn in. This is very important since you will need this information in order to make sure that your file is imported properly (dimension-wise) into SketchUp. This is discussed more in **Part 1—Importing of AutoCAD Files** exercise.

Importing Issues with AutoCAD Files

There are two fundamental issues that you, as a modeler, have to be aware of when importing AutoCAD files into SketchUp for further modeling work. The first main issue, which a lot of modelers seem to come across, is that content is imported with lines that are not fully continuous. The second issue is connected to elevation discrepancies, when two-dimensional content is not imported as flattened but rather has an elevation component connected to a portion of geometry. This section is dedicated to the review of these issues and to provide options on how to best solve them.

As mentioned in the introduction to this section, the first main issue is connected with content being imported from AutoCAD where lines are not fully continuous. One way to notice that the imported file might be suffering from this issue is the lack of created surfaces in SketchUp, either at a small section of the drawing or on the entire drawing. The reason for this issue is that certain lines will not intersect each other when they are imported from AutoCAD into SketchUp, thus preventing SketchUp from creating surfaces. The issue can be seen only if you zoom far enough into an area where you are suspecting a break. Besides the inability to create a surface, this particular issue can give you incorrect dimensions when measured or prevent you from creating right angles at that particular intersection, because you are not selecting the correct intersection point.

One way to overcome this issue is to engulf the imported drawing with a rectangular surface utilizing the **Rectangular** tool. After this step is completed, locate true intersections with the help of assist lines generated by the **Tape Measure** tool and redraw the lines with the **Line** tool. Another option is to use a third-party application that analyzes your imported drawing and corrects the areas of incomplete lines.

The second issue is created when two-dimensional AutoCAD files are imported into SketchUp, but they are not truly two-dimensional in nature. If the AutoCAD file that is imported into SketchUp was not previously flattened, meaning all the elevation geometry is set to zero, then most likely you will have some modeling issue down the line if the imported geometry has an elevation component. One option to prevent this issue is to flatten the AutoCAD file prior to importing it into SketchUp. Use the AutoCAD command **FLATTEN** to change the elevation component of the geometry to zero by typing it into the command line, selecting the geometry, and pressing the **Enter** button.

Importing AutoCAD Files to SketchUp

The following SketchUp tutorial will shows a detailed workflow on how to properly import, set up, and use your AutoCAD drawings into SketchUp. The tutorial is split into two parts. Part 1 goes over a series of steps in order to properly import the AutoCAD file. Part 2 dives deeper into how to set up the imported file and properly use it to create three-dimensional objects. For the tutorial, you will need an AutoCAD file in order to follow the example. The Example in **Fig. 3.2** is an AutoCAD file containing a drawing of a typical 32-in. barrier wall to be imported to SketchUp.

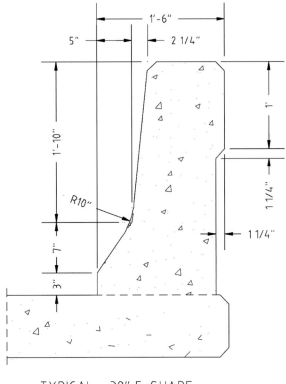

TYPICAL - 32" F-SHAPE

Figure 3.2   Content of the AutoCAD file being prepared for import into SketchUp.

Part 1—Importing of AutoCAD Files

Before you begin with the workflow, you need to enable Scenes, Entity Info, and Layers inspectors as follows:

1. From the **Main Menu**, with a left mouse click, select the **Window** menu and then select the **Scenes, Entity Info,** and **Layers** (**Fig. 3.3**). In SketchUp version 2020, the **Layers** inspector window was renamed to **Tags**; it is the exact same inspector but with a different name assigned to it. You will have to repeat the process for each selection, since the **Window** menu closes as soon as you make your selection.

When step 1 is completed you can start working on the actual file importing portion of the exercise—in your case, to import the AutoCAD file of your choosing.

Figure 3.3   Example on how to activate the **Entity Info, Layer,** and **Scenes** inspectors.

2.   From the **Main Menu**, with a left mouse click, select **File** then **Import…**—this will open the **Import** window (**Fig. 3.4**).

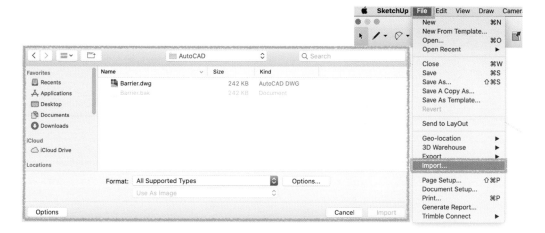

Figure 3.4   Location of the **Import** option and the file selection window.

The next step is to select the type of file you would like to import. Technically you can leave the **Format** selection drop-down menu as is, since by default it is set to **All Supported Types**, but in order to be consistent with the example, select the file format you plan on using.

3. From the **Import** window, left click on the **Format** drop-down menu and select **AutoCAD Files (*.dwg, *.dxf)** (see **Fig. 3.5**).

Figure 3.5   List of supported files for import by SketchUp.

4. Navigate to the location of your file, for example, "**Barrier.dwg**" AutoCAD file, and click the actual file but do not click the **Import** button yet. Click on the **Options…** button that is located next to the **Format** drop-down menu (**Fig. 3.6**) When the

Figure 3.6   **DWG Import Options** dialog box and the **Scale** drop-down menu.

Options... button is selected (**1**), a submenu **DWG Import Options** will appear (**2**) (see **Fig. 3.6**). Under the **Scale** subsection (**3**), click on the **Units** drop-down section and choose the proper units that coincide with the unit of the AutoCAD file. This is particularly important since it will prevent SketchUp from "guessing" the units and importing the drawing with incorrect dimensions.

Modeling Tip

Besides the **Scale** subsection option, there are two additional subsections present:

- **Position** subsection: By selecting the **Preserve drawing origin**, SketchUp will locate your AutoCAD drawing at the same spatial location. This option is very important for site modeling, locating substructure components, and importing survey data.
- **Geometry** subsection: The options in this subsection are more tailored when importing three-dimensional drawing.
  - **Merge coplanar faces** option helps as to how polygon faces are imported. When importing polygon faces, SketchUp has a tendency to import them as triangulated, regardless of whether they reside on the same plane. When this option is selected, all the coplanar faces are not triangulated but rather imported as polygon faces.
  - **Orient faces consistently** option will cause all faces to be imported in the same direction, therefore eliminating a problem when some faces are imported reversed.
  - **Import materials** option is selected by default, and will be helpful when importing three-dimensional models that have materials associated with them.

Press the **OK** button in order to save your choices from the **DWG Import Options** menu and click the **Import** button.

---

5. After the import had been completed, SketchUp will provide you with the **Import Results** window. The **Import Results** window shows you a list of what was imported (Entities Imported) and what was ignored (Entities Ignored) from the drawing in terms of layers, blocks, arcs, circles, and so on. Click on the **Close** button to move forward to Step 6 (**Fig. 3.7**).

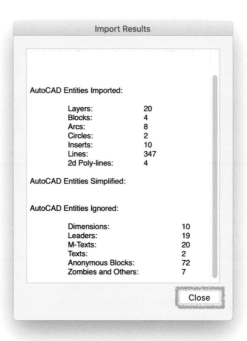

Figure 3.7 **Import Results** window opened automatically after the AutoCAD file was imported.

6. Our next step is to check the geometric accuracy of the drawing that was imported into SketchUp with the help of the **Tape Measure** tool. In order to check the geometric accuracy of the imported file, you will want to measure a representative geometric distance on the imported drawing. A geometric distance can be any portion of the imported drawing that can be easily verified, for example, width, length, or height of a beam, column, concrete deck, steel girder, and so on. For this example, the **Tape Measure** tool will be utilized to measure the height of the barrier wall, as shown in **Fig. 3.8**. If the measured distance corresponds to the distance specified in the drawing, you can move to the next step. In the case where the measured distance does not correspond to the specified distance, type the new distance in the **Measurements Input Box** and press **Enter**. This will resize the imported drawing by keeping the newly typed value as a constant and resizing the rest of the imported drawing.

Modeling Tip

The resizing of the imported drawing by utilizing the **Tape Measure** tool can be done multiple times by repeating step 6 of the workflow.

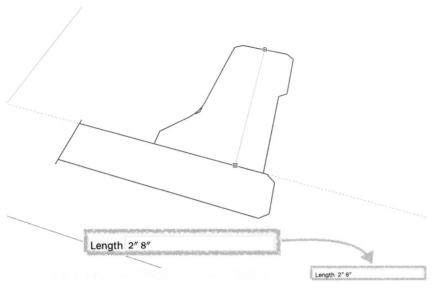

Length 2″ 8″

Length 2″ 8″

Figure 3.8   Example of choking the accuracy of the imported content.

Up to this point in the example, the workflow reviewed how to import the drawing (part 1 of the example) and also how to ensure that the imported drawing keeps the correct geometric accuracy. Checking the dimensions of an imported drawing in SketchUp should become second nature to any modeler. This is one of the most important portions of this workflow because the imported drawing is the base to your three-dimensional model, and if the base is incorrect, everything that is modeled and layered becomes incorrect. Setting up a correct base will save you a lot of productive downtime that can be associated with redrawing, scaling, and remodeling.

Layers (Tags), Scenes, and Proper Setup of AutoCAD Files in SketchUp

The next set of steps will prepare the already-imported drawing for three-dimensional modeling use. As mentioned at the beginning of this chapter, the imported drawing should only be used as a base or tracing layer for your three-dimensional model and not be part of what you are modeling. As such, with this idea in mind, the next several steps will achieve this steps will achieve this goal.

Layers (**Tags** in SketchUp version 2020), **Scenes**, and **Entity** Info dialog boxes are looked at first before continuing to the exercise.

Layers (Tags) Inspector

The **Layers** inspector allows you to manage individual layers in your drawing (**Fig. 3.9**).

1. The plus sign (+) allows you to add a new layer.
2. The minus sign (−) allows you to delete a layer.
3. The **Eye** icon allows you to toggle the visibility of all edges and surfaces that are currently present under the layer.

4. The Name tab shows the names of layers present in your model. The name of the layer can be changed by clicking over the name portion.

5. The Dashes tab allows you to change the line type of all the edges that are present in the layer.

6. The **Pencil** icon shows the default layer under which all the edges and surfaces will be located as you model. The pencil icon can be reassigned to different layers just by clicking on the box next to the layer you want to reassign it to.

7. The color box option allows you to color-code different surfaces that are part of your three-dimensional model after you assign specific colors to the layers you want to be color coded. The color-coded option is activated by choosing the arrow icon and left-clicking the **Color by Layer** option on the submenu (**Fig. 3.9**). This option is very useful when modeling different components of bridges (substructure vs. superstructure components), rebar, structural steel, tendons and tendon equipment, and the like. These topics are covered in more detail throughout the book.

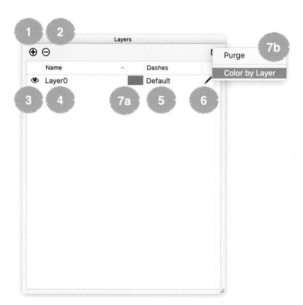

Figure 3.9   **Layers inspector window.**

Entity Info Inspector

The **Entity Info** inspector allows you to inspect entities that are present in your model and also manipulate/reassign layers to the selected entities (**Fig. 3.10**).

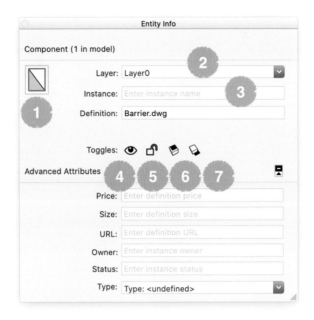

Figure 3.10   **Entity Info** inspector window.

A quick overview of all the options under the **Entity Info** inspector is as follows:

1. The rectangle option box, located in the upper-left side of the inspector, allows you to change the color of the faces that are associated under a specific layer.

2. The **Layer** drop-down menu allows you to reassign layers to selected entities.

3. The **Instance** and the **Type** options are more advanced than we need for now.

4. The **Eye** icon toggles the visibility of your entity, whether it is three-dimensional or two-dimensional in nature.

5. The **Lock** icon will lock your entity in place. When a lock entity is selected, all the lines that are associated with that entity will be colored in red.

6. The **Shadow** 1 icon will prevent your entity from receiving shadows when the **Shadows** tool is activated. This is particularly useful for visualizations and presentation; Chapter 12-Introductions to Scenes, Section Cuts, Shadows, and Fog.

7. The **Shadow 2** Icon, when selected prevents your entity from casting shadows when the **Shadows** tool is activated. Similarly to point 6, this option is particularly useful for visualization and presentation purposes; **Chapter 12-Introductions to Scenes, Section Cuts, Shadows**, and Fog.

8. The **Advanced Attributes** section of the **Entity Info** dialog box is explained in more detail in **Chapter 4—Introduction to Information Modeling and Organization**.

Scenes Inspector

The **Scenes** inspector is the location where you can manage your scenes that are present in you three-dimensional model (**Fig. 3.11**).

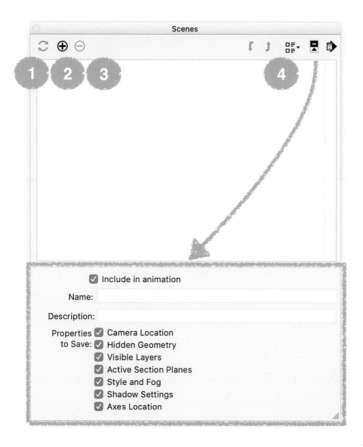

Figure 3.11   **Scenes** inspector window.

1. The **Update** icon that looks similar to a recycle sign allows you to update/save any changes to all or a specific scene.

2. The plus sign (+) allows you to add a new scene.

3. The minus sign (−) allows you to remove one or multiple scenes.

4. The five icons to the right of the **Scene** Inspector are self-explanatory. When the **Show/Hide Detail** icon, point 4, **Fig. 3.11** is pressed, it will show a total of eight attributes. These attributes are connected to an individual scene and will change certain properties/options of the scene. Since they impact different options, they are discussed in more detail in **Chapter 12—Introduction to Scenes, Section Cuts, Shadow, and Fog**

Part 2—Proper Setup of AutoCAD Files

1. The first step is to create your base layer by utilizing the **Layers** inspector. Click on the plus (+) sign, which will create a new layer. Without clicking anywhere else,

change the new layer name to "**1BaseLayer.**" The addition of the numeral 1 will automatically position the newly created layer below the default layer, which is **Layer0** (**Fig. 3.12**).

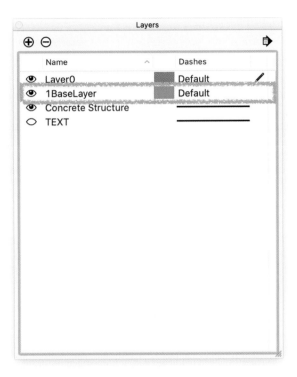

Figure 3.12   The **Layers** inspector and the newly created **1BaseLayer** layer.

Modeling Tip

In SketchUp version 2020, the **Layer0** is renamed to **Untagged**. Besides the changes in nomenclature, there are no additional changes to other options.

2. The next step is to move all the layers that are automatically imported from the AutoCAD file to your newly created layer, **1BaseLayer**. Select the imported drawing by clicking on it once. When the drawing is selected, in the **Entity Info** inspector, click on the **Layer** drop-down box and select the **1BaseLayer**. This step will move all the graphical components that are imported into SketchUp to your new layer (**Fig. 3.13**).

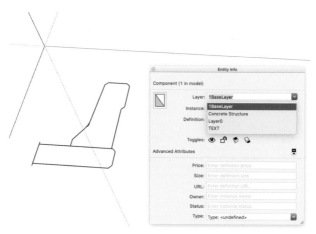

Figure 3.13   Moving the imported geometry to the new layer, utilizing the **Entity Info** inspector window.

3. When step 2 is completed, select all the layers, except for **1BaseLayer** and **Layer0** which is your default layer, and click on the minus (–) sign. A secondary window will appear with three options. Choose "**Move to Current Layer**," and press the **Delete** button. The content of the selected layers will not be deleted. It will only be moved to the default layer, but the layer entity will be deleted and removed from the **Layers** inspector window (**Fig. 3.14**).

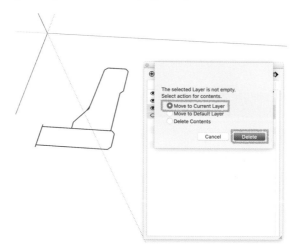

Figure 3.14   Removing layers and moving content to the current, default layer.

Modeling Tip

The same result will be obtained if you choose "**Move to Default Layer**" since your current layer is also your default layer. Please remember that this is not always true. In situations where you manually assign the "pencil" icon to a different layer aside from **Layer0**, **Layer0** no longer becomes your current layer.

Modeling Tip

The last option on the secondary window list is "**Delete Contents**." By clicking on it, you will delete your imported drawing. The "**Delete Contents**" option can be used for content that was imported into SketchUp but will not be needed in your model. Use this option with reservation, since some of the layers are intertwined between the drawings you need and drawing you do not need in the model; therefore, the possibility of deleting something that you will need in your base drawing is high. Follow the steps outlined under section **AutoCAD Files, Issues, and Interface**, and prepare the AutoCAD file rather than manipulate the content in SketchUp.

4. In order to check if step 3 was completed correctly, press the **Eye** icon located next to **1BaseLayer** under the **Layers** inspector. This operation will toggle the visibility, and the base drawing should disappear from the screen. Pressing the **Eye** icon again will make the base drawing appear again.

5. The next step is to create two scenes where you can toggle the visibility of the base drawing. From the **Scenes** inspector, press the plus (+) sign, which will generate a scene. Select the generated scene and rename it with **BaseLayerON**, which can be found under the **Name:** tab. Also deselect all the checked selections except for **Visible Layer**. Click on the update arrows, and select **Update** from the secondary window (**Fig. 3.15**).

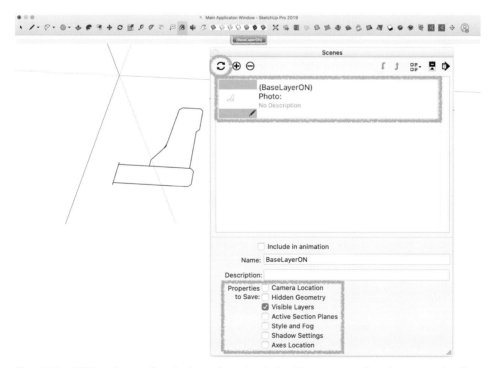

Figure 3.15   Addition of a scene from the **Scenes** inspector window. The scene created in order to create the effect where the visibility of the base drawing can be switched **ON** or **OFF**.

6. Repeat the same process as outlined in step 5 for your second layer but with a slight difference. Create the second layer by pressing on the plus (+) sign and rename it **BaseLayerOFF**. From the **Layers** inspector click on the **Eye** icon next to the **1BaseLayer**; this will toggle the visibility of the layer, and the base drawing will disappear. Click on the update arrow found under the **Scenes** inspector, and select the **Update** button from the secondary window (**Fig. 3.16**).

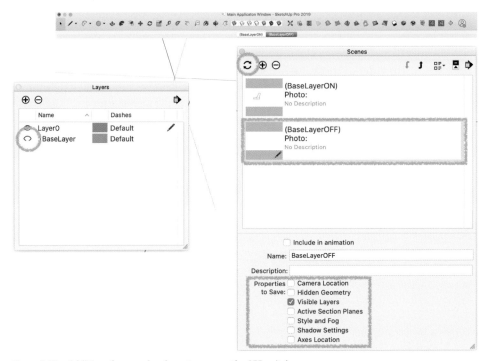

Figure 3.16   Addition of a secondary layer to serve as the OFF switch.

You are now done with your AutoCAD file setup, and you are ready to move toward three-dimensional modeling! The following is an explanation of what was just done in the last two steps of the exercise. The imported drawing should be used only as a base drawing (as a guide or a template) and should not be a part of the three-dimensional drawing, similarly as you would use tracing paper in a real-life application. When the **BaseLayerOFF** scene is selected, the imported drawing will disappear. When the **BaseLayerON** scene is selected, the imported drawing will reappear. The ability to toggle the visibility on and off of the imported drawing is tremendous, since it will allow you to check on your three-dimensional modeling progress as you draw on top of your base drawing. Another positive outcome of applying this workflow is the ability to delete the base drawing from your model when you are done with it and thus shrink your SketchUp file size. The base file can be deleted by deleting the **1BaseLayer** in the **Layers** inspector. The process is similar to what was done in step 3, with one difference; you will select the "**Delete Contents**" from the secondary window as an option.

CHAPTER 4

Introduction to Information Modeling and Organization

Information modeling and the proper organization of three-dimensional models represents the basic concepts or ideas that this book is based upon. From personal experience, the SketchUp application represents a very versatile platform that can offer a variety of different types of information, besides just the visual type. With proper organization of components and the model in general, a plethora of information can be shared and used, in real time, by estimators, engineers, schedulers, project managers, field personnel, owners, and so on.

The following section was developed to introduce you to a universal workflow on how to properly and efficiently organize your models. Furthermore, this section will also strengthen your understanding of how to access and use the different types of available information that is generated by properly following the organization workflow. Unlike other chapters, this chapter concentrates less on modeling work and more on pure organizational logic.

The development of a detailed model as shown in **Fig. 4.1** should start at the initial stages of a construction project, when the project is estimated.

The estimating stage of a project represents the very first time that the contractor becomes familiar with the different aspects of the project and also with any possible flaws. Furthermore, besides this initial introduction to the project, the contractor also has to prepare a detailed list of material quantities. What if you could combine these two steps together?

Now you can go one step further. After the estimating team is successful in the pursuit, all the bid material will be transferred to the field management team. The field management team will have to reanalyze and understand the different aspects of the data and make an additional effort to reshape it for their own purposes. The logical question is how to streamline and remove inefficiencies from the process. The core workflow idea of this book is meant to accomplish this, through proper organization of your modeling work in a way that it will be beneficial for each team in the construction industry.

In order for the three-dimensional model to be equally useful for the estimating and the construction team, you should follow a series of steps on how to organize your model efficiently for this purpose:

- **Review the Contract Drawings:** Prior to modeling, you have to review and separate the contract drawings per individual structures. For example, when you are in the process of a pier cap review, combine all the drawings that are associated with that particular pier cap (piles, footing, reinforcement bars, drainage or fire suppression systems, formwork, bearings, post-tensioning, etc.). In addition, you also have to pay attention to key components, such as the overall geometry of the structure (cross

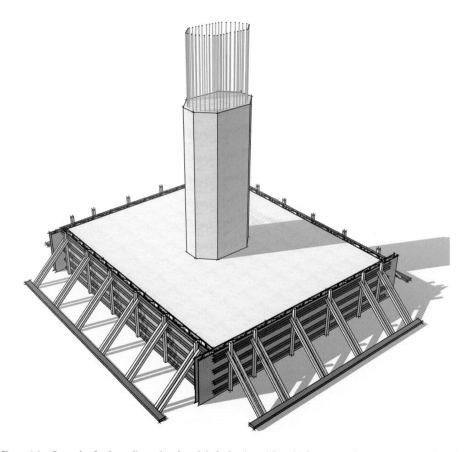

Figure 4.1 Example of a three-dimensional model of a footing with multiple construction components activated.

slopes, break lines, etc.); architectural details, if any; and the location of construction joints, if shown. In some cases where construction joints are not shown, review the constructibility of the structure and assign your own construction joints.

- **Use of Layers:** Layers should be created and assigned for each separate component in the structure, for example, reinforcement bars, drainage pipes, piles, and the like. In addition, these layers should be further separated by each construction section of the structure, where the components are located, for example, reinforcement bars in the first stage of concrete placement versus reinforcement bars in the second stage of concrete placement. This is a particularly important step because you can show and hide certain sections of the model based on the construction schedule and also calculate quantities for that portion of estimation or construction work.

- **Use of Components:** Use of components is a must for any type of model where a large number of three-dimensional objects are repeated. An example of this rule is a situation dealing with reinforcement bars, piles, pipes, and so on. Components allow you to have a greater control over different parts of the structure, since SketchUp's "stickiness" property will not apply for the edges and faces. Furthermore, components give you the ability to fully replace or partially revise structural objects without the need to delete and reinstall them in the model. The workflow steps for this property are covered in detail in **Chapter 1—Components and Groups**. From the information side, the use of components can generate an automatic count for the quantity of each component used in the model. This SketchUp property is discussed later in this chapter. Finally, the use of components will keep the overall size of your model file significantly lower, in contrast, if you use multiple copies of groups or just have one continuous model. Decreasing the file size of your model will increase the overall response time of SketchUp and make the file management and transfer task easier to accomplish.

- **Accurate Modeling of Component Interface:** Special attention should be paid to modeling components that intersect each other in the structure. An example where this requirement will apply is cases where piles and the pile cap intersect each other or the intersection between a pile cap and post-tensioning ducts. This is particularly important in order to acquire the correct information with regard to the volume of the structure. Not having the correct volume information can impact the overall estimate of material quantities, among other things.

The next step is to review how you can apply the previously mentioned organization points to a three-dimensional model. For this purpose the three-dimensional model shown in **Fig. 4.1** will be used. The pier footing and column consists of HP14 × 141 piles, as supports, pier footing, and a column. Furthermore, the pier footing is separated into two different placement areas: one for pier footing itself and one for the column. Other details included in the three-dimensional model are the reinforcement bars and the overall formwork.

Our first stop is to review the **Layers** inspection window (**Fig. 4.2**).

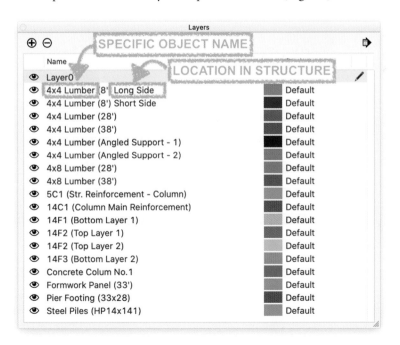

Figure 4.2   List of layers depicting different three-dimensional components that are part of the pier footing model.

As you can see by observing the inspection window, individual layers have been created for each object present in the pier footing structure. Furthermore, the layers have the following nomenclature: (**Specific Object name _ Location in Structure**). The **Specific Object Name** portion of the nomenclature describes just that, for what type of object the placeholder was created (concrete column, pier footing, piles, reinforcement bars, formwork lumber, etc.). The **Location in Structure** portion of the nomenclature describes which construction stage that particular object belongs to. For example, you will want to differentiate between a particular reinforcement bar located in the pier column and the pier footing. As shown in **Fig. 4.2**, the pier footing has multiple copies of the same #14 longitudinal-type bar. In order to designate in which section of the footing it is located, multiple layers with the same prefix and different suffix were created—see the last portion of the nomenclature shown in the **Layers** inspection window (**Tags** in SketchUp version **2020**) (**Fig. 4.2**).

Another organization point that you can see in the **Layers** inspection window (**Tags** in SketchUp version 2020) is the color blocks (**Fig. 4.3**).

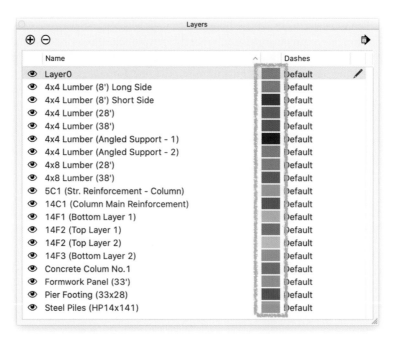

Figure 4.3   Different color codes are assigned to different layer types in the **Layers** inspection window.

Each layer is separated and organized by a different color. From construction and three-dimensional modeling experience, choose colors that are opposite to each other for objects that are located in the same vicinity (**Fig. 4.3**). You can choose any colors that fit your particular style, but remember to make them easily distinguishable from each other. When you are happy with the color organization, you can activate this option by selecting the **Color by Layer** property located in the **Layers** inspection window (**Tags** in SketchUp version **2020**) (**Fig. 4.4**). The **Color by Layer** property is highly helpful during estimating reviews and also for the creation of work plans during construction.

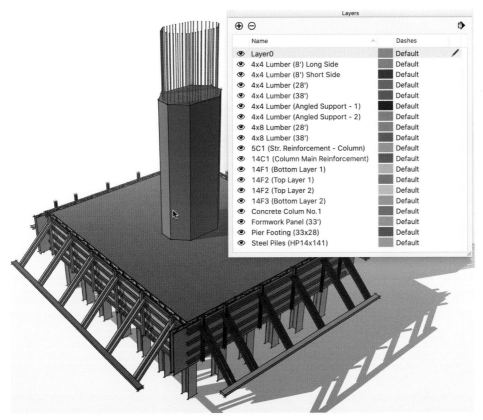

Figure 4.4   Layers on the three-dimensional model are colored by the preselected color codes.

The last point concerning the **Layers** inspection window (**Tags** in SketchUp version 2020) is the ability to toggle the overall visibility **ON** and **OFF** for individual components. This is particularly helpful when you want to concentrate on one portion of the structure. For example, in **Fig. 4.5** the column concrete mass together with the concrete column reinforcement and the footing concrete mass were hidden from view.

By hiding all the unnecessary components, the estimating or construction team can focus on only the area where they want to check the estimated time against the scheduled time or review it with the field personnel for the amount of work present in one shift. The same procedure of turning the layer visibility **ON** or **OFF** can be used when producing construction tracking drawings. With small changes to the color coding or the layers, the drawing can depict what was completed for a particular time allocated in the schedule versus what is planned in the next segment of allocated schedule time (**Fig. 4.5**).

The next step is to review what type of information attributes are present in the **Entity Info** inspection window and how you can prepare it for future use. All the information

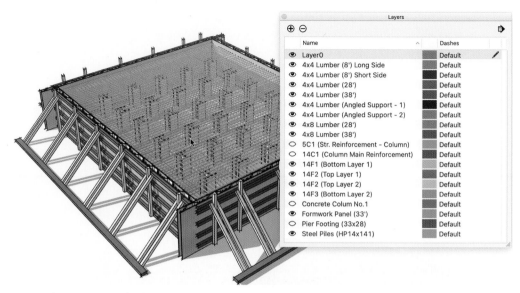

Figure 4.5   Example where the concrete outer layer of the column and the accompanied reinforcement bars are hidden by turning the visibility off.

presented in the **Entity Info** is inputted by the user and therefore can be updated as the modeling work progresses. **Figure 4.6** shows the list of information attributes generated when a single HP14 × 141 pile is selected:

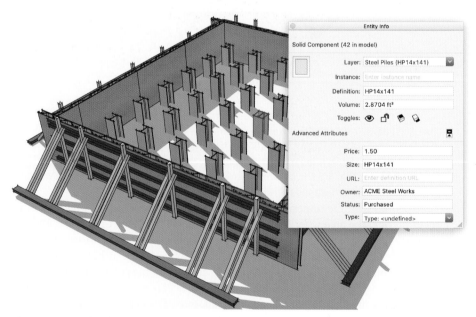

Figure 4.6   Example of pile quantity shown in the **Entity Info** inspection window.

1. **Number of solid components: Entity Info** calculates the total number of copies that are present for the selected pile type. Only the same pile type that is part of a particular footing will be counted by the **Entity Info** window in a group.

2. **Layer**: Specifies the name of the layer that the pile stirrup belongs to.

3. **Definition**: Represents the name of the component that was chosen when the component was created.

4. **Volume**: Reveals the total volume of one pile. This can be converted into weight by multiplying by the correct material density.

5. **Price**: Reveals the current price of the pile. This can be updated as the estimate work progresses or if new prices are allocated during construction.

6. **Size**: Reveals an easily recognizable and very specific size of a component. For example, reinforcement bar size, pile size, pipe diameter, strand diameter, and so on are all viable types of information that can be entered with regard to the size of the component information.

7. **Owner**: Reveals the company who produces or sells the material. This can be especially important for estimating and construction purposes.

8. **Status**: The status section can be used to denote if the item was procured and the specific date of procurement as shown in the example. This can be highly useful during estimate work or when cost analysis is performed during the construction phase.

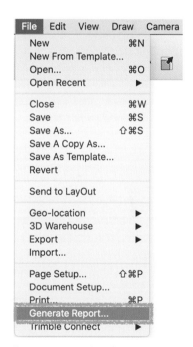

Figure 4.7   Example on how to access the **Generate Report...** option from the main menu.

The preceding list of information attributes represents a snapshot for one component of many that are a part of the three-dimensional model. A very useful tool in SketchUp, which will take your model to the next level in information modeling, is the ability to generate a report, containing all eight information attribute fields, for all the components present in the three-dimensional model. The following steps expand on the process in order to generate a report:

1. The first step in the process is to activate the **Generate Report** inspection window. Select **File** from the main menu, and then select the **Generate Report...** option (**Fig. 4.7**).

2. The type of information that you want to extract from your model can be tailored from the **Generate Report** inspection window shown in **Fig. 4.8.**

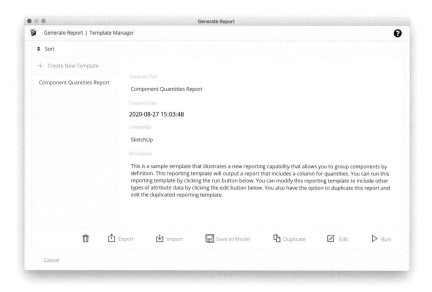

Figure 4.8 **The main section of the Generate Report window allows you to assign a name, date, and other information to your report.**

3. Select the **Edit** option located on the bottom portion of the **Generate Report** inspection window (**Fig. 4.9**).

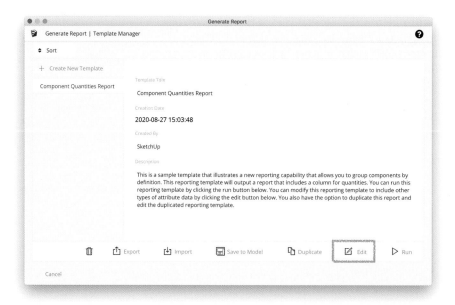

Figure 4.9 **The Edit option in the Generate Report window allows you to further tailor all the information that you wish to be present in your report.**

4. From the **Model Attributes** column (**a**) relocate the attributes you want to include in the report to the **Report Attributes** column (**b**) by selecting and clicking on the arrow key (**c**), (**Fig. 4.10**)—the eight attributes listed earlier in this chapter is a good starting point.

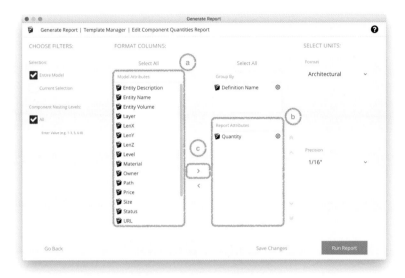

Figure 4.10   Example of how to add specific attributes to your report.

5. Aside from the ability to add desired attributes in the report, SketchUp also gives you the ability to limit the area for the generated information to the entire model or a selection. In addition, you can also select the format of your units and the precision, which usually is associated with dimensional attributes (**Fig. 4.11**).

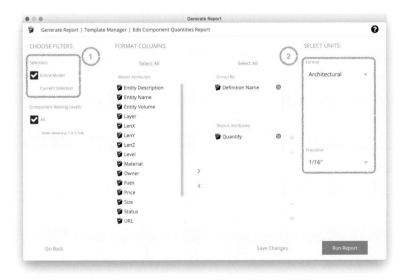

Figure 4.11   Example of additional attributes that can be modified in the Generate Report window.

6. The "gear" icon located next to the attributes allows the user to edit the name of a specific attribute and also adjust the rules on how the information is generated (**Fig. 4.12**).

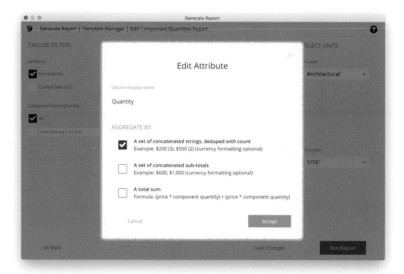

7. Click on the **Run Report** button in order to complete the process and to generate the report (**Fig. 4.13**).

Figure 4.13 Final view of all the changes made to the **General Report** window, prior to selecting the **Run Report** button.

8. By clicking on the **Run Report** button, SketchUp will generate a preview of all the data and all the attributes selected in the previous step. If you are dissatisfied with the content, click on the **Go Back** button to make further adjustments; otherwise, click on the **Download** button (**Fig. 4.14**). The file format used to save the list of attributes is ***.csv**, which can be opened by any application that supports spreadsheet-type information.

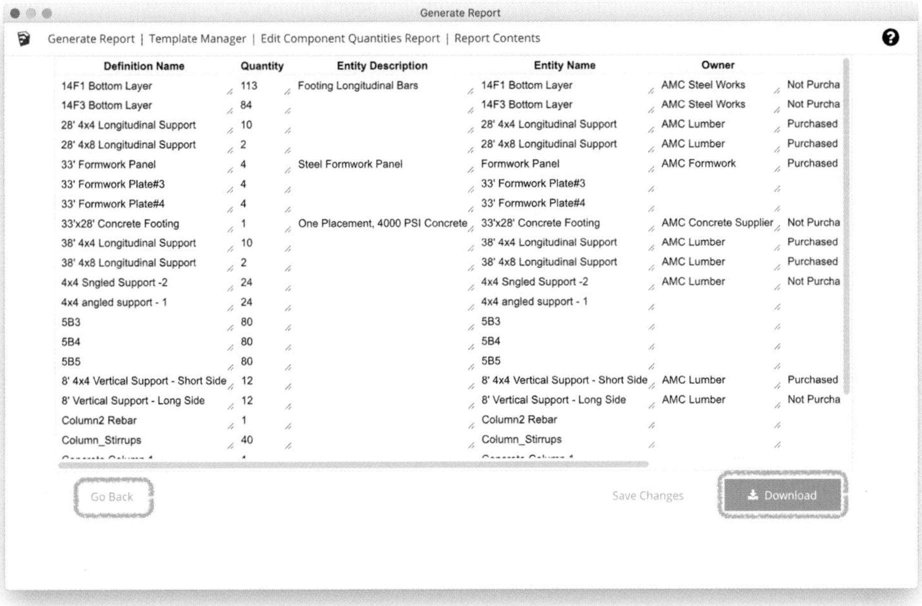

Definition Name	Quantity	Entity Description	Entity Name	Owner	
14F1 Bottom Layer	113	Footing Longitudinal Bars	14F1 Bottom Layer	AMC Steel Works	Not Purcha
14F3 Bottom Layer	84		14F3 Bottom Layer	AMC Steel Works	Not Purcha
28' 4x4 Longitudinal Support	10		28' 4x4 Longitudinal Support	AMC Lumber	Purchased
28' 4x8 Longitudinal Support	2		28' 4x8 Longitudinal Support	AMC Lumber	Purchased
33' Formwork Panel	4	Steel Formwork Panel	Formwork Panel	AMC Formwork	Purchased
33' Formwork Plate#3	4		33' Formwork Plate#3		
33' Formwork Plate#4	4		33' Formwork Plate#4		
33'x28' Concrete Footing	1	One Placement, 4000 PSI Concrete	33'x28' Concrete Footing	AMC Concrete Supplier	Not Purcha
38' 4x4 Longitudinal Support	10		38' 4x4 Longitudinal Support	AMC Lumber	Purchased
38' 4x8 Longitudinal Support	2		38' 4x8 Longitudinal Support	AMC Lumber	Purchased
4x4 Sngled Support -2	24		4x4 Sngled Support -2	AMC Lumber	Not Purcha
4x4 angled support - 1	24		4x4 angled support - 1		
5B3	80		5B3		
5B4	80		5B4		
5B5	80		5B5		
8' 4x4 Vertical Support - Short Side	12		8' 4x4 Vertical Support - Short Side	AMC Lumber	Purchased
8' Vertical Support - Long Side	12		8' Vertical Support - Long Side	AMC Lumber	Not Purcha
Column2 Rebar	1		Column2 Rebar		
Column_Stirrups	40		Column_Stirrups		

Figure 4.14 **The Generate Report window allows you to review the actual report prior to be downloaded as a *. csv file type. Additional changes can be made by selecting the Go Back button.**

The report, in the spreadsheet-type format, can be further tailored by adding columns in order to generate subtotal prices per material component, total prices per section of the project, notes—that can be used for estimating or project management purposes, and so on. Furthermore, the generated report can be easily shared among team members during the estimating or construction phases of the project or even among subcontractors who are on the way to become new members of your team. The options are truly endless, but proper organization as described in this chapter is a must in order to be successful.

Imagine the overall benefits for a company that can implement the information modeling workflow, covered in this book, on a daily basis. Benefits in terms of increased accuracy during estimating, increased safety and quality during construction, and an increase in the overall communication quality between the contractor, subcontractors, and the owner— technically the heavy civil project will be virtually constructed long before the construction team is physically on location.

The idea of information modeling, organization, and the implementation procedures is further covered through a set of step-by-step workflow examples in the next set of chapters, in the book. Each of these chapters concentrates on a specific portion or structure of a heavy civil project, with the intent for you to gain experience in implementing the core ideas of this book, which are information modeling and proper organization.

CHAPTER 5

Modeling of Substructure Components

Pier Footing
Copy
 Array Copy

Solids Toolset
Union and Subtract tools
Volume Estimating

The main idea behind this chapter is to review the benefits that three-dimensional models can provide for estimating purposes or during construction of a project. With proper planning and execution, these benefits can extend beyond just the visual aspect into quantity information and scheduling. For this purpose, the modeling workflow for pier footings is reviewed as well as the organization process of the components and proper planning.

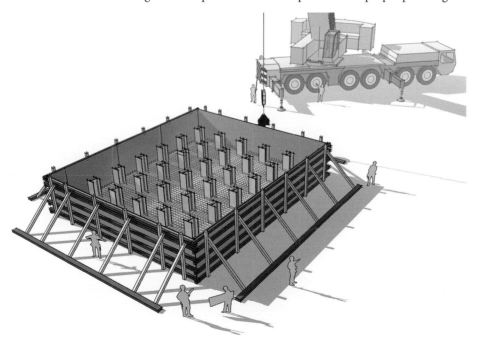

Pier Footing

The first step is to model the piles. For this example HP16×141 will be used (**Fig. 5.1**). The HP16×141 pile is doubly symmetrical in nature, and therefore, you will only model one-quarter in order to save time—for more information, see **Chapter 8—Modeling of Girders**.

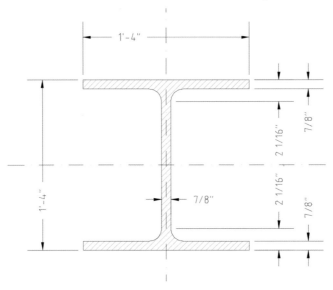

Figure 5.1 **Representation of HP16×141 pile.**

1. Select the **Rectangle** tool and create a surface with the following dimensions 8 in. × 8 in. (**Fig. 5.2**).

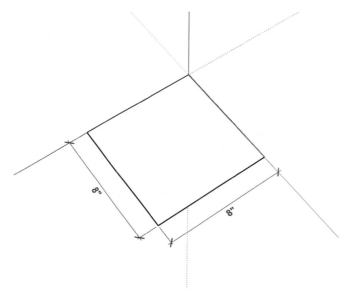

Figure 5.2 **Creation of an 8-in. × 8-in. surface that will be used as a base to create the cross-sectional view of the pile.**

2. Select the **Tape Measure** tool, and place one vertical assist line, indicating the edge of the web and two horizontal assist lines indicating the thickness of the flange and fillet (**Fig. 5.3**). When placing the assist lines, follow the dimensions as shown in **Fig. 5.1**.

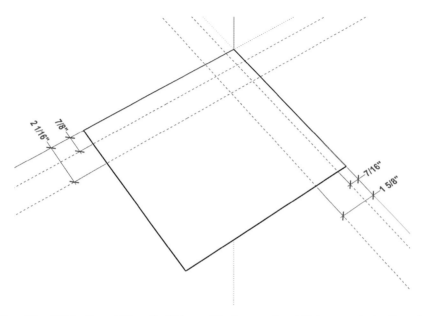

Figure 5.3 Utilizing the assist lines, the thickness of the flange, web, and fillet were marked on the surface.

3. Utilizing the **Line** and **Arc** tools and trace the outline of the quarter pile (**Fig. 5.4**). Select and erase the unnecessary surface present outside of the traced pile (**Fig. 5.4**).

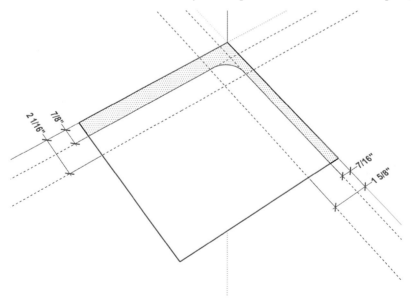

Figure 5.4 Outline of one-quarter of the pile.

4. Utilizing the **Move** tool, make an offset copy of the quarter pile. Select the quarter pile and while holding down the **Option** key on your keyboard, click and drag an offset copy of the quarter pile (**Fig. 5.5**).

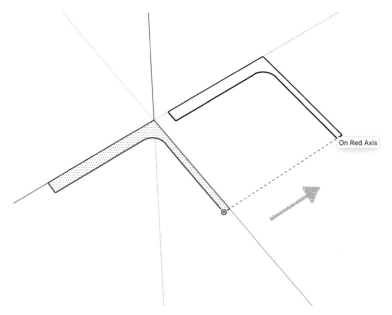

On Red Axis

Figure 5.5   An initial offset copy is created in order to complete one-half of the pile.

5. Select the offset copy and utilizing the **Scale** tool, mirror the offset copy by left-clicking on the middle green box and dragging it across until the value in the **Measurements Input Box** shows a −1.0 (**Fig. 5.6**). You can also type in the −1.0 value in the **Measurements Input Box**, which will automatically create the mirror.

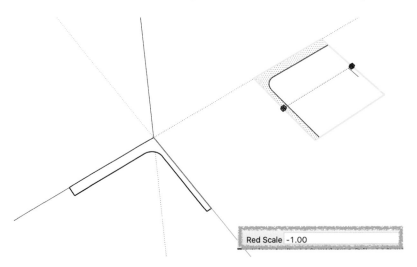

Red Scale -1.00

Figure 5.6   The copied section is mirrors utilizing the **Scale** tool.

6. Select the mirrored copy of the quarter pile and utilizing the **Move** tool, left-click on the inner edge and combine it with the other quarter half of the pile (**Fig. 5.7**). Erase the middle line—congratulations you have created one half of a pile (**Fig. 5.7**).

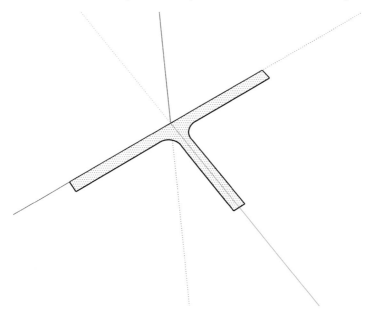

Figure 5.7   The two quarter segments are joined together to form the first half of the pile.

7. Repeat steps 5 throughout 7 to complete the entire pile (**Fig. 5.8**).

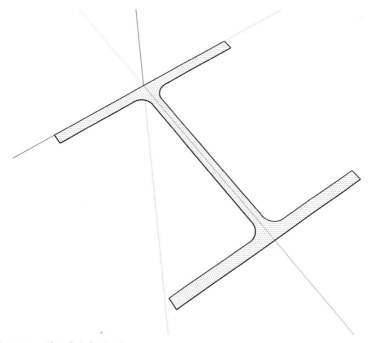

Figure 5.8   The pile is finalized.

8. Extrude the pile for an arbitrary 10 ft., utilizing the **Push/Pull** tool (**Fig. 5.9**).

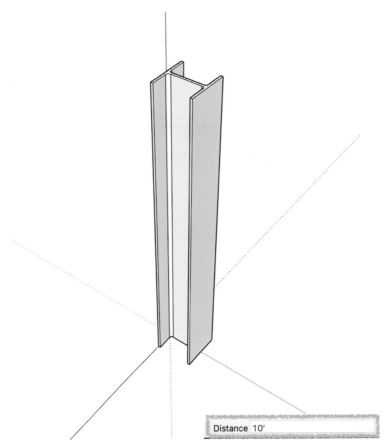

Distance 10'

Figure 5.9   Extrusion of the pile is done utilizing the **Push/Pull** tool.

9. After the pile is created, it is time to create a component from the pile and create a pile group. Select the entire pile, with a right mouse click anywhere on the pile open the context menu. From the context menu, select **Make Component...** (**Figure 5.10**). In the **Definition** field, write the pile size, in this case **HP16×141**. In the **Description** field, write the location of the pile, for this example write **Pier Footing 1**. Under the **Advanced Attributes** section, write an arbitrary price of 1.50. The usage of the **Advanced Attributes** fields is discussed toward the end of this chapter and was also discussed in **Chapter 4** in more detail (**Fig. 5.1**). Press the **Create** button to complete the process.

10. The next step is to create a layer for the pile. From the **Layers** inspection window (**Tags** in SketchUp version 2020), click on the plus (+) icon. For the name of the

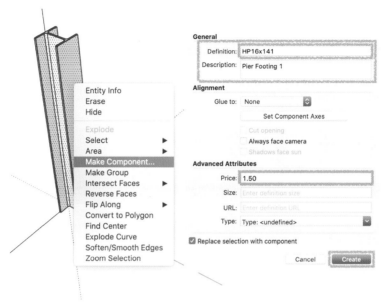

Figure 5.10   Converting the three-dimensional pile into a component for ease of use.

layer use **Pier Footing 1 Pile** (**Fig. 5.11**). Select the pile and from the **Entity Info** inspection window, click on the **Layers** drop-down menu and select the **Pier Footing 1 Pile** layer (**Fig. 5.11**). You moved the entire geometry of the pile to the created layer.

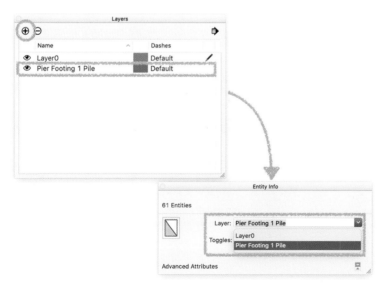

Figure 5.11   Creating a new layer in the **Layers** inspector window and transferring the pile geometry to the new layer utilizing the **Entity Info** inspector window.

11. The pile group will be spaced as per **Fig. 5.12**.

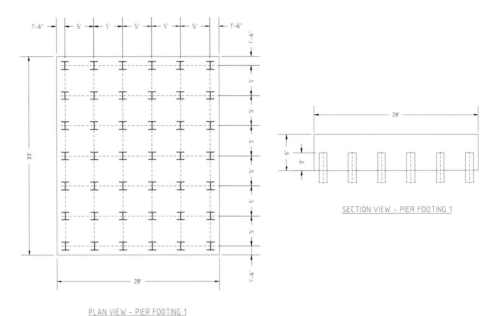

Figure 5.12   Sketch of the pile plan and pile elevation that will be used for example purposes.

Now make an offset copy of the first pile for a total distance of 5 ft by utilizing the **Move** tool. Select the pile and then select the **Move** tool. Hold down the **Option** key and click on any corner on the pile. Drag the copy perpendicular to the pile for a total distance of 5 ft (**Fig. 5.13**). Type the 5 ft distance in the **Measurements Input Box** and press **Enter**.

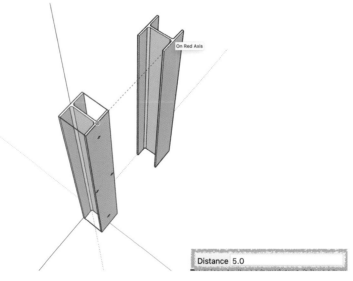

Figure 5.13   Creation of an offset copy of the pile component.

12. When the first offset copy is completed, type **5X** in the **Measurements Input Box** and press **Enter** (**Fig. 5.14**). SketchUp will automatically create the additional four pile copies needed for the pile group.

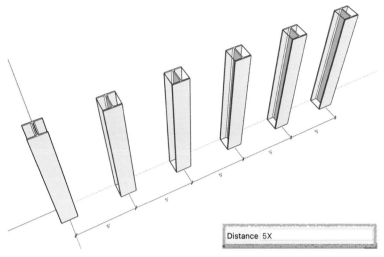

Distance 5X

Figure 5.14   Utilizing the array copy option in order to make multiple copies of piles from a single entity, at a predetermined distance. The array copy option saves you precious modeling time.

13. Select the entire pile line that you created in step 12. Repeat step 11, in order to make an offset copy. Similarly to step 11, create a parallel offset copy of the pile line with a spacing of 5 ft (**Fig. 5.15**).

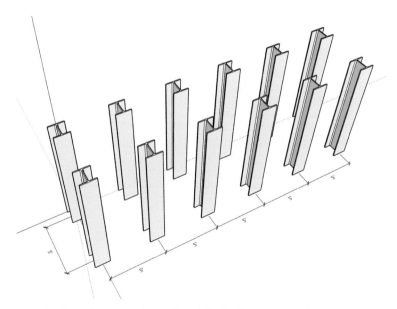

Figure 5.15   Copying and preparing the next line of piles for the array copy option.

14. When the offset copy is completed, type **6X** in the **Measurements Input Box** and press **Enter** (**Fig. 5.16**). SketchUp will automatically create the additional five pile line copies required for the pile group.

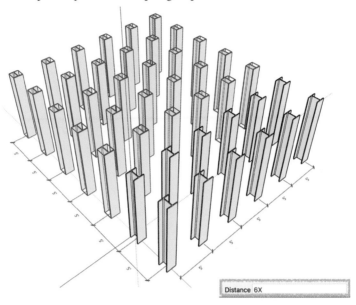

Distance 6X

Figure 5.16   Utilizing the array copy option to create the rest of the piles.

15. Utilizing the **Tape Measure** tool, create two assist lines as shown in **Fig. 5.17**. Offset the assist lines for a distance of 1.5 ft (**Fig. 5.17**). You created your first corner of the pier footing.

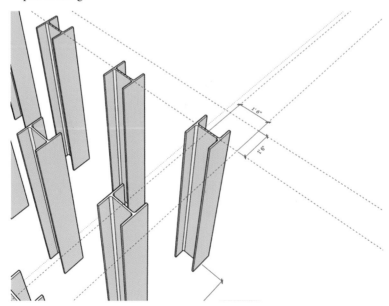

Figure 5.17   Creation of the first footing corner, utilizing the Tape Measure tool and assist lines.

16. Select the **Rectangle** tool, and utilizing the intersection of the offset assist lines as a starting point, create a surface with the dimensions of 28 ft ×33 ft (**Fig. 5.18**).

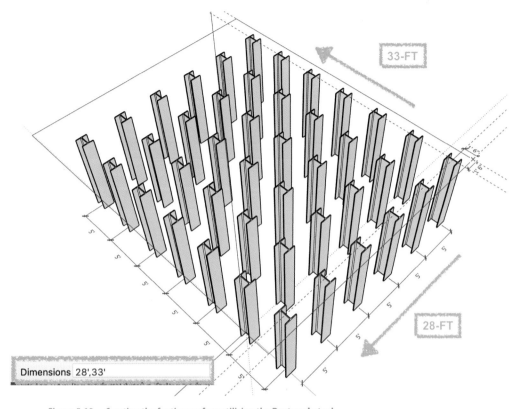

Figure 5.18   Creating the footing surface utilizing the **Rectangle** tool.

17. Select the rectangular surface created in Step 16, and with the help of the **Move** tool, lower the surface 3 ft on the blue axis. (If you notice in **Fig. 5.12**, the piles are embedded for a total length of 3 ft into the footing.) Utilizing the **Push/Pull** tool, extrude the thickness of the footing, which for our example is 6 ft (**Fig. 5.19**).

18. The next step is to convert the pier footing into a component. Select the three-dimensional rectangle, without the piles, and activate the context menu by a right mouse click anywhere on the object. Select the **Make Component...** option (**Fig. 5.20**). In the **Definition** field, write the pile footing name, in this case use **Pier Footing 1**. In the **Description** field, write the location of the pile, for the purpose of this example write **Pier Footing 1** (**Fig. 5.20**). Press the **Create** button to complete the process.

19. With the new component created, it is time to create a separate layer for the footing. From the **Layers** inspection window (**Tags** in SketchUp version 2020), click on the plus (+) icon. For the name of the layer use **PF1 Placement 1** (**Fig. 5.21**). Select the

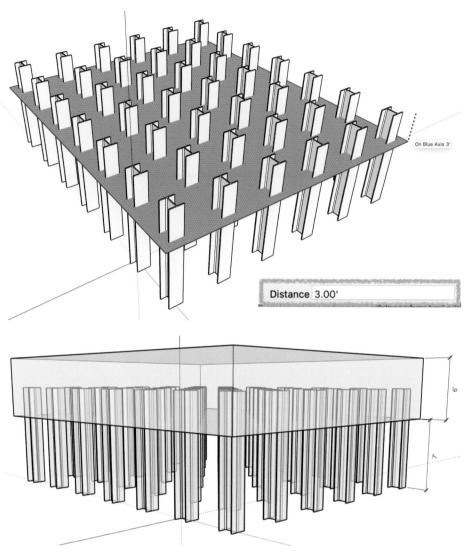

On Blue Axis 3'

Distance 3.00'

Figure 5.19 Locating the rectangular surface to the correct elevation on the footing and extruding the footing utilizing the **Push/Pull** tool.

pile, and from the **Entity Info** inspection window, click on the **Layers** drop-down menu and select the **PF1 Placement 1** layer (**Fig. 5.21**).

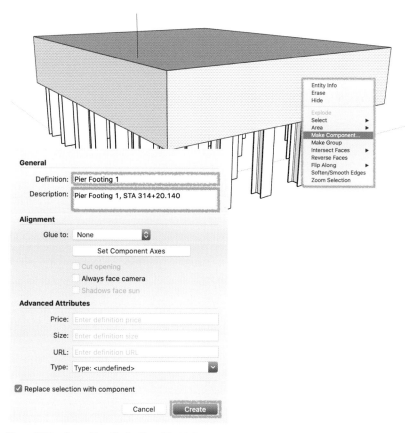

Figure 5.20   Converting the footing object into a component.

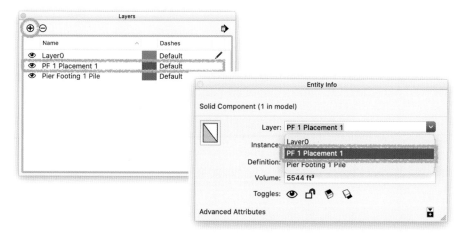

Figure 5.21   Creating a layer in the Layers inspector window for the footing component and transferring the geometry to the new layer utilizing the Entity Info inspector window.

In the next set of steps the **Union** and **Subtract** tools will be utilized in order to subtract the volume of the steel piles from the volume of the concrete footing (**Fig. 5.22**).

The **Union** and **Subtract** tools are part of the **Solids** toolset. You can only subtract one solid from another, in this case that will be one pile at a time from the pile footing. This would not be an issue if you only had one or two piles per footing. Now imagine if you have a bridge with sixty piles per footing and a total of twelve piers, each having two footings per pier, the amount of clicking increases exponentially, and thus the time to complete the task. For that reason use the **Union** tool, which will create a single union of a solid from our group of piles. By creating the union, you can subtract all forty-two piles from the pile footing with one selection.

The subtraction of the piles from the footing will give you an accurate estimate of the volume and thus the amount of concrete in the footing itself. This can be valuable information for estimating purposes or for material purchases during construction activities.

20. The first step is to select all the piles and make a copy of them by either (1) selecting **Edit** and then **Copy** from the menu bar or (2) pressing the **Command** and **C** keys at the same time on your keyboard. Do not paste the copied piles at this moment.

21. Select the piles once more, if they got deselected, and click on the **Union** tool. This will create a union of all the selected piles.

22. The next step is to subtract the piles from the footing. Select the **Subtract** tool and hover over the piles. SketchUp will alter your mouse pointer to add a numeral to it. Click on the piles and then click on the footing. The piles will disappear, and you will be left with impressions of the piles into the footing (**Fig. 5.23**).

23. The last step is to paste the copied piles from Step 21 either by selecting **Edit** then **Paste** from the menu bar or by pressing the **Command** and **V** keys at the same time on your keyboard. Use the pile impressions in the footing as a guide for placement (**Fig. 5.24**).

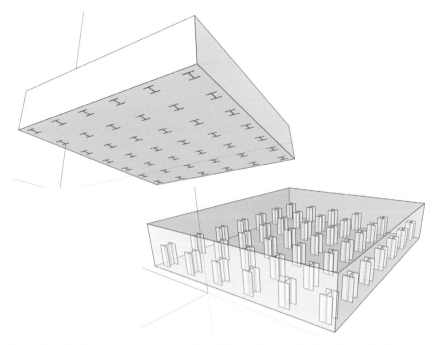

Figure 5.23   The **Subtract** tool creates an outline of all the piles embedded into the pile footing.

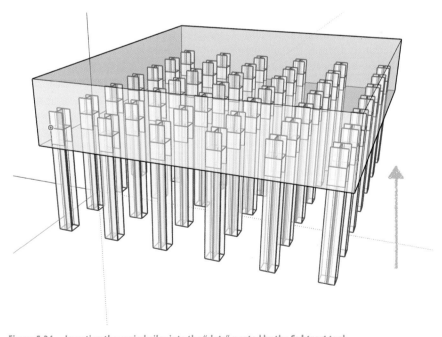

Figure 5.24   Inserting the copied piles into the "slots" created by the **Subtract** tool.

Step 23 completes this exercise. The exercise provides the modeler or engineer with two very useful tools, a good visual representation of the footing, including the piles, and spatial information in regards to the structural components. Either of these tools can be highly useful on an individual basis or as a group for estimating purposes or construction work.

If you turn to the **Entity Info** inspection window, and select the footing component, SketchUp will provide you with the overall volume of the footing (the footing volume excludes the individual volume of the piles, the reason for the extra work done in Steps 21 steps 20 through 23 of the exercise). **Figure 5.25** shows a side-by-side volume comparison prior to and after the embedment of the piles into the footing mass.

Figure 5.25   Visual change in the volume prior to and after the embedment of the piles into the footing mass.

The volume can be easily converted into cubic yards which can be used for estimating purposes or for procurement of concrete material during construction. Granted the footing used in this exercise has a very simple geometry, but you have to look at the bigger picture when the footing has a more complex geometry and the chances of making a mistake in the overall estimate increase.

The other reason for creating individual components besides spatial information is the ability to receive quantity information. If you select any of the piles and you turn your attention to the **Entity Info** inspection window, in the upper-left corner SketchUp will provide the number of that specific component present in the model (**Fig. 5.26**).

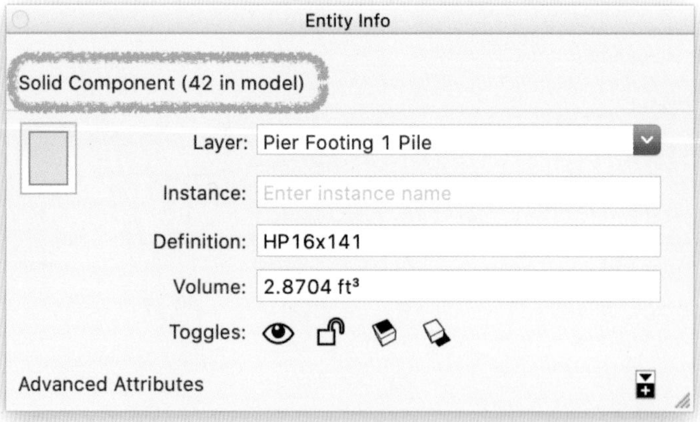

Figure 5.26   Quantity provided by the Entity Info inspector window on the number of pile present in the footing.

With the proper grouping and setup, as shown in the preceding exercise, an engineer or estimator can keep a very accurate count on number of different components needed for a specific structural part of the project, in this case the footing.

Granted this example only has two different components, but imagine the possibility if you add all the separate components and create a fully functional model of the footing as shown in **Fig. 5.27**.

The model shown in **Fig. 5.27** is composed of the piles, footing rebar, column rebar, footing concrete, and column concrete (shown to the first construction joint). Each of these components are separated into individual layers and color-coded accordingly for ease of review by the teams, regardless of whether the project is in the estimating stage or construction stage.

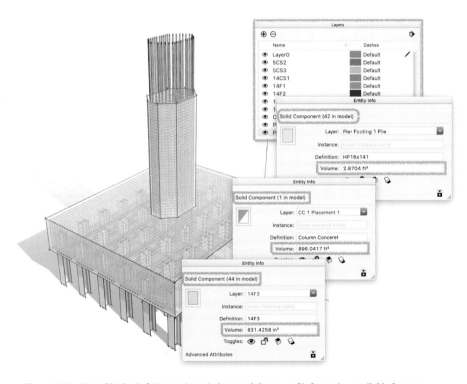

Figure 5.27   List of **Entity Info** inspection windows and the type of information available for usage.

Modeling of Steel Reinforcement

There are many factors, such as design, aesthetics, type of use, and cost efficiency, just to name a few, which are responsible for the increase in the overall complexity of structural components on a project. This increase in complexity can also create installation challenges during construction because traditional two-dimensional drawings become inefficient in conveying all the necessary information. These challenges can also cost the contractor valuable time and resources and at the same time create a lower-quality product for the owner.

One of the immediate benefits of having a three-dimensional drawing is an overall increase in the ability to better visualize the components, be that on paper or electronically, and the ability to perform clash detection by importing different components in the same drawing (**Fig. 6.1**).

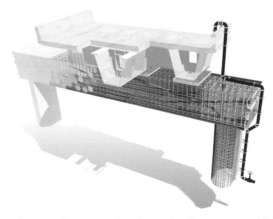

Figure 6.1   Three-dimensional representation of a complex pier structure modeled in SketchUp.

103

One of the areas that by far has the largest construction challenges from the increase of structural complexity is the installation of reinforcement bars. By creating three-dimensional models of the steel reinforcement in structural components, you are also making a virtual construction site, where mistakes can be "planned out" before they become timely and costly during construction.

This chapter deals with turning the traditional two-dimensional reinforcement drawings, which could be construction or shop drawing, into three-dimensional models. This chapter starts with the creation of one of the more complex bar bends, which is the stirrup; continues with the proper way to create a highly useful component and how to group multiple components; and finishes with the creation of a steel-reinforced element that is reviewed later in this chapter. At the very end of the chapter there is a list of modeling tips that will make your three-dimensional modeling experience easier and will eliminate some recurring mistakes.

Reinforcement Bar Sizes

When working with reinforcement bars, it is a good idea to review how individual bars are sized. There are a total of eleven different size bars currently used in construction and they range from a #3 to #18. Modern reinforcement bars have outside deformation, unlike their earlier counterparts that were completely smooth. The outside deformation is created in order for the bars to interact better with the concrete. The reinforcement bars are nominal and they are measured to the outside of the deformation. See **Fig. 6.2a** for a typical representation of a **Reinforcement Bar** and the associated **Bar Size** versus **Diameter at Deformation** correlation.

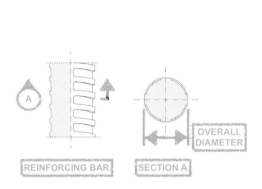

Outside Bar Diameter

Bar Size	Approximate Diameter Outside Deformations (in)
#3	7/16
#4	9/16
#5	11/16
#6	7/8
#7	1
#8	1 1/8
#9	1 1/4
#10	1 7/16
#11	1 5/8
#14	1 7/8
#18	2 1/2

Figure 6.2 (*a*)   Typical reinforcement bar dimensions.

Note that the use of the correct outside diameter in three-dimensional modeling is very important, especially when working with structures where the reinforcement bars will protrude through a steel plate girder, as shown in **Fig. 6.2b**, or when accounting for the clear cover especially in reinforced concrete caps, beams, or columns.

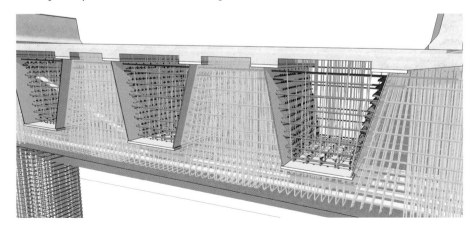

Figure 6.2 (*b*)   Example of reinforcement bars protruding a steel girder.

Stirrup Bend Modeling

The stirrup bend type, besides structural performance, is also responsible for the creation of the outer boundary or shell of the steel-reinforced concrete component. This chapter is separated into three sections. The first section examines the workflow of creating an accurate representation of a three-dimensional stirrup type bend. The second section examines the procedure of preparing the created stirrup object for later use in the steel-reinforced concrete model. The third section examines the procedure of pairing reinforcement bars and creating components of specific groups.

Section 1—Stirrup Modeling

There are several types of stirrups that are most commonly used in construction projects. Stirrups can be separated based on the overall shape (rectangular, circular, or helical), the number of legs (two-legged, four-legged, or six-legged) or whether they are a closed or an open type (**Fig. 6.3**).

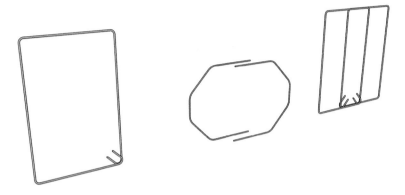

Figure 6.3 **Different stirrup bend types modeled in three dimensions.**

For this example the most common type stirrup in use is modeled, which is the rectangular, two-legged, closed stirrup. The stirrup is made from a #5 (5/8-in. diameter) reinforcement bar with hook length of 10-in. and a bar bend diameter of 2-1/2 in. (**Fig. 6.4**).

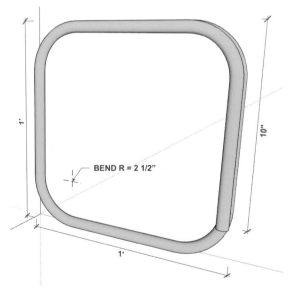

Figure 6.4 **Geometric properties of the stirrup.**

The procedure to model this stirrup is as follows:

1. Start by drawing the centerline of the stirrup shape, utilizing the **Rectangle** tool. For this example arbitrary dimensions of 12 in. × 12 in. were chosen (the dimensions are measured to the outside of the stirrup shape, so you will have to subtract one-half thickness of the rebar). Since you are using a #5 reinforcement bar, you will subtract 0.3125 in. from each side, or 5/8 in. total (**Fig. 6.5**).

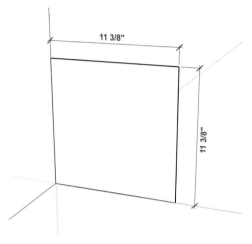

Figure 6.5   A 11 3/8-in. x 11 3/8-in. surfce is created in order to start the modeling proceudre of the stirrup. The surce represents the center-line of the stirrup.

2. After the bar centerline is drawn, next you will find the center of the stirrup by drawing assist lines with the help of the **Tape Measure** tool. This is a very important step in the workflow; finding the center of the stirrup will help you with the accurate placement in the reinforced concrete object. With the **Tape Measuring** tool, clicking on one side of the stirrup and using SketchUp's inference midpoint, find the first center. Repeat this step for the second midpoint and the second center (**Fig. 6.6**).

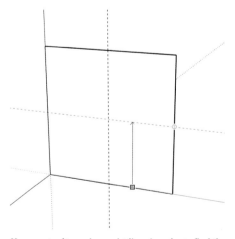

Figure 6.6   Utilizing the **Tape Measure** tool to assign assist lines in order to find the center points of the surface.

3. The next step is to create the bend diameter of the stirrup. First use the **Tape Measure** tool to create assist lines by selecting the centerline at each side of the stirrup and offsetting the assist line by 2-1/2 in. type the offset distance in the **Measurements Input Box** (**Fig. 6.7**).

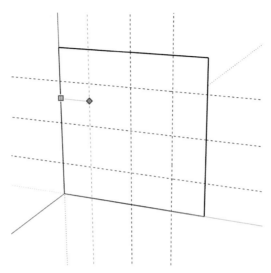

Figure 6.7   Utilizing the assist lines to find the start and end point of the bar bend diameter.

After all the assist lines are created, the second step is to create the bend diameter by using the **2 Point Arc** tool. Left-click on each intersection point between the assist line and the bar centerline and draw an arc that is tangent with the bar centerline—the inference engine will help you to snap on the points and find the tangent (the line will change color to magenta when it is tangent; (**Fig. 6.8a**). When completed, delete the unnecessary lines past the created arcs (**Fig. 6.8b**).

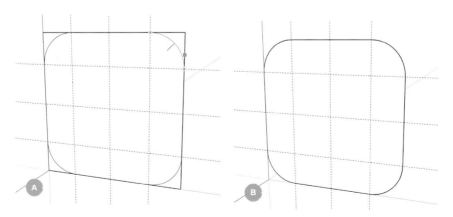

Figure 6.8 (a)   The arc turns to magenta color when it is tangent to the stirrup center line. (b) Completed view of the stirrup.

4. Select one arc corner and a side. Select the **Move** tool, hold down the **Option** key on your keyboard and make a copy of the selected lines, confirming the copied lines are parallel to the original set. Offset the copied lines by one diameter of the reinforcing bar, which is 5/8 in. (**Fig. 6.9**). This is the first step in creating stirrup hooks.

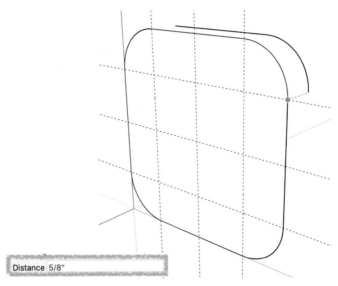

Distance 5/8"

Figure 6.9   Offsetting a copy of a line to create the stirrup hook.

5. The next step is to delete the connecting line between the two corner arcs, which are located across the lines you copied in step 4 (**Fig. 6.10**).

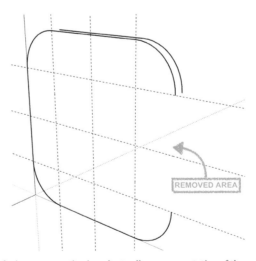

REMOVED AREA

Figure 6.10   Removal of the inner-connecting in order to allow proper rotation of the outside geometry.

6. Because of the nature of the stirrup bend, you will have to rotate the corner arcs in order to make the proper connection between the original and copied components. Select the **Rotate** tool and left-click at the base of the corner arc, confirming the rotate command is perpendicular to the corner, by pressing the corresponding arrow key (top arrow—Blue axis, left arrow—Green, right arrow—Red axis). Rotate the corner arc until the assist line of the **Rotate** tool coincides with the base point of the arc that was offset in step 4. Repeat the same step for the offset arc but in reverse, making certain that the assist line coincides with the base point (**Fig. 6.11**).

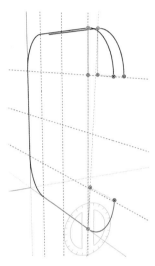

Figure 6.11   Rotation of the external geometry in order to connect to the stirrup hook.

7. Utilizing assist lines as guides, draw two edges, angled as the rest of the geometry, in that section of the stirrup (**Fig. 6.12**). Delete all the assist lines prior to proceeding to Step 8.

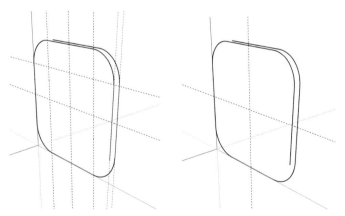

Figure 6.12   Last step in the creation of the stirrup hook extensions, assist lines are used to mark the correct length and provide guidance on the direction of the line. The second drawing is the representation of the completed stirrup, including the hooks.

8. The next step is to draw the diameter of the reinforcing bar. Using the **Circle** tool, click on the end of the centerline (either centerline end will do), making certain the circle is parallel to the centerline. Make use of the SketchUp's inference engine in order to properly position the circle and press the corresponding arrow keys on your keyboard to properly align it to the correct axis. (top arrow—Blue axis, left arrow—Green, right arrow—Red axis). You do not have to modify the standard number of sides, but keep in mind that for more complex structures that have a high number of bars you will have to reduce the number of edges (**Fig. 6.13**).

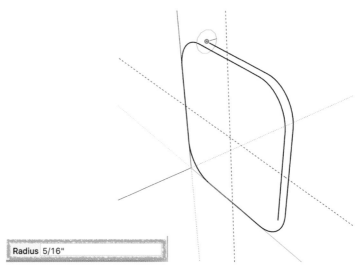

Radius 5/16"

Figure 6.13   Creation of the diameter of the reinforcement bar.

9. The last step is to use the **Follow Me** tool to create the stirrup. Select the face of the circle that you created in step 8, and with the help of the centerline, extrude the bar (**Fig. 6.14**).

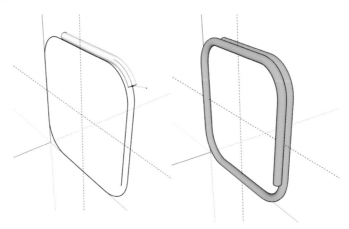

Figure 6.14   The overall thickness of the stirrup was provided utilizing the **Follow Me** tool and the circle (bar diameter) created in step 8.

Section 2—Create a Stirrup Component

This section explores the workflow of what to do with the stirrup component that you created in the first section of this chapter. Steel-reinforced concrete objects have hundreds and in some cases thousands of different types of reinforcement bars intertwined tightly. Properly naming these bars, making them into components, and setting up layers are highly important in order to keep track and create a successful three-dimensional model. When you configure a layer and make a component of a specific reinforced bar, you give yourself the tools to properly and easily manage that particular bar in the reinforced concrete object. Also you give yourself the ability to toggle the visibility on and off in order to increase the system response time when you model the structure. Additionaly, the ability to toggle the visibility on and off of the reinforcement, after the model is completed, allows you and your team to better visualize and understand the overall reliathionship of the reinforcement and other components in the structure. The following workflow can be equally applied to large structures that are highly complex and to small structures with less complexity.

The first step is to create a component of the three-dimensional stirrup model:

1. Select the entire stirrup. If the assist lines that show the center of the stirrup were selected also, deselect them by holding down the **Shift** button and clicking on them.

2. With the right mouse button, click anywhere on the selected stirrup and from the context menu choose **Make Component... (Fig. 6.15)**.

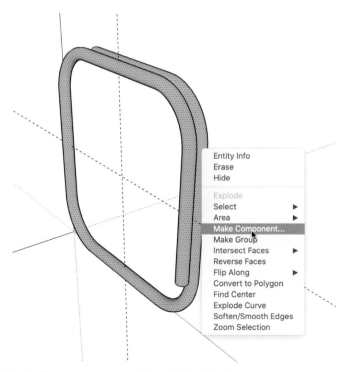

Figure 6.15   Activating the context menu in order to select the Make Component... option.

3. At the component inspection window, give the proper name and description of the stirrup and also set the local axis. It is recommended to name the models of the reinforced bars by the tag names that were assigned either in the shop drawings or the construction drawings. This is an important step in order to properly tag and later identify the specific reinforcement in the drawing. For this example, give the stirrup component an arbitrary name of **1S5** (Item No, description of the bar, and size of the bar). In the description box, add notes to the specific grade of the reinforced bar and the outside finish (epoxy coated, black or stainless steel). Under the **Advanced Attributes** section, populate the fields **Price** ($1.50—this is an arbitrary price per pound of steel) and **Size** (#5—the actual size of the reinforcement bar). The use of the **Advanced Attributes** information was explained in more detail in **Chapter 4— Introduction to Information Modeling and Organization** (**Fig. 6.16**).

Figure 6.16   Creating a component from the stirrup-type bar and adding information to different attribute boxes.

4. The next step is to position the local axis of the stirrup objects. For this purpose use the assist line that you created in the first part of the exercise as a location placeholder. From the component inspection window, click on the **Set Component Axis** button.

The mouse cursor will change from an arrow to the axis symbol. Set the axis cursor in the center of the assist lines and press the mouse button three times in order to set the *x*, *y*, and *z* axis, respectively (**Fig. 6.17**).

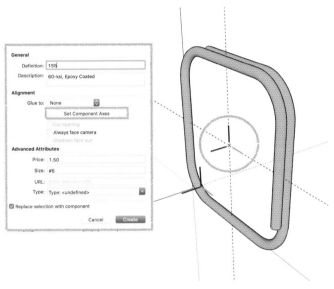

Figure 6.17   Creating a new local axis for the reinforcement bar. The center location was chosen on purpose in order to assist with later placing the bar.

5. After the local axes are set on the stirrup object, click on the **Create** button, and the component will be created (**Fig. 6.18**).

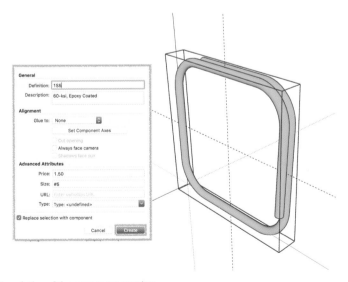

Figure 6.18   Completion of the component creation.

Modeling Tip

The position of the local axis should be based on the type and physical location of the reinforced bars within the reinforced concrete object that you are modeling. For the earlier example, the stirrup represents a horizontal type bar; therefore, logically the local axis should be set at the center of the stirrup. The reason for this is the ability that it gives you to position the stirrup in the center of the reinforced concrete object. If you are dealing with the main longitudinal bars, you would have positioned the local axis on the horizontal face of the reinforcement bar. Similarly for ties, the local axis will be positioned where the longitudinal bar would pass through.

The next step is to create a layer that will partner with the newly created stirrup component:

1. From the **Layers** inspector window (**Tags** in SketchUp version 2020), add a new layer by clicking on the (+) plus sign. Give the newly created layer the same name as the component, **1S5**, see step 3 (**Fig. 6.19**).

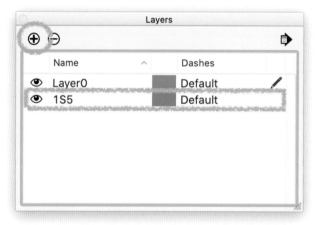

Figure 6.19   Creation of a new layer (1S5) in the **Layers** inspector window for the new stirrup reinforcement bar.

2. In addition to the name, also assign a color by clicking on the color box and specifying a color (**Fig. 6.20**). The color assignment will help you visually identify the reinforcement bar by activating the **Color By Layer** option located in the **Layers** inspector window.

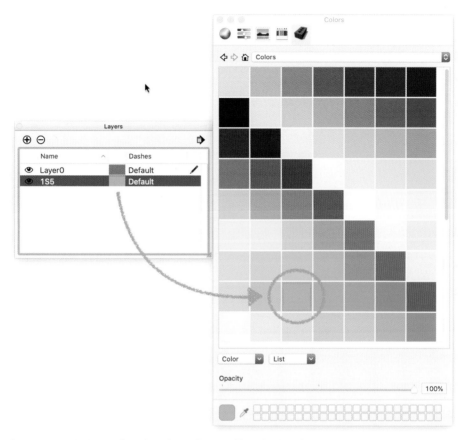

Figure 6.20   Assigning a color code to the newly created layer for the **Color By Layer** option of visual identity.

Modeling Tip

Make the color assignment based on the location of the rebar. For example, is the rebar located in a column, cap, deck, beam, and so on or is it associated with a group of longitudinal or transverse bars? Another concept to remember is to assign contrasting colors, for example, assign a light green color to longitudinal bars and pair them with a peach color for the main vertical reinforcement bars, the stirrups. **Figure 6.21** shows a list of colors and their assignments. In order to achieve a similar visual impact as shown in **Fig. 6.21**, all of the concrete surfaces were assigned to a **ConcSurf** layer, with a color assignment of white indicating very low **Opacity** (in the range of 10–15 points).

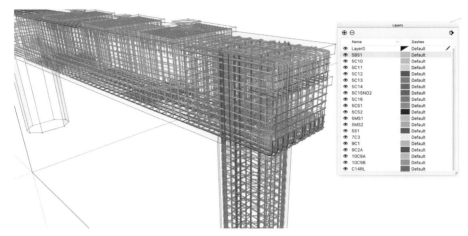

Figure 6.21   Visual example of a pier cap of how the Color By Layer option works in complex three-dimensional models. As seen in the figure, the Color By Layer enhances the visual representation of each type of reinforcement present in the structure.

3. When you are done with the options in the **Layers** inspector window (**Tags** in SketchUp version 2020), you will turn your attention to the **Entity Info** inspection window, in order to assign the newly created component to the correct layer. Select the stirrup component, and click on the **Layer** drop-down menu. From the **Layer** drop-down menu, change the default layer to the newly created layer, in our case **1S5** (**Fig. 6.22**).

Figure 6.22   Moving the rebar geometry to the newly created layer (1S5) with the help of the Entity Info inspector window.

4. In order to check if the correct layer was assigned to the stirrup component, select the **1S5** component and toggle the visibility (eye icon) **OFF** from the **Layers** inspector window (**Tags** in SketchUp version 2020); the component will disappear from the screen (**Fig. 6.23**).

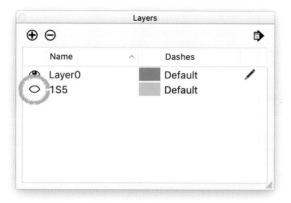

Figure 6.23   Ability to activate and deactivate the visibility of a component from the **Layers** inspector window (**Tags** in SketchUp version 2020).

Section 3—Pairing of Reinforcement Bars

This section of the **Stirrup Modeling** workflow examples examines the procedure of pairing reinforcement bars and creating specific groups. For this example you will need to create a stirrup-type bar and a tie-type bar. Follow the previous steps on creating a reinforcement bar. You can use data from a previous project or imagine something on your own.

Pairing of reinforcement bars is a common practice in construction, and it is done for many reasons. In this section, we are not concerned with the reasons for pairing but with the procedure of making the rebar pairs easy to work with and at the same time adding attributes for easy visual recognition. **Figure 6.24** shows one section of a pier cap with very tight pairing of reinforcement. Without proper pairing and the addition of attributes to each of the reinforcement bars, modeling a pier cap as shown in **Fig. 6.24** can be very difficult and time-consuming.

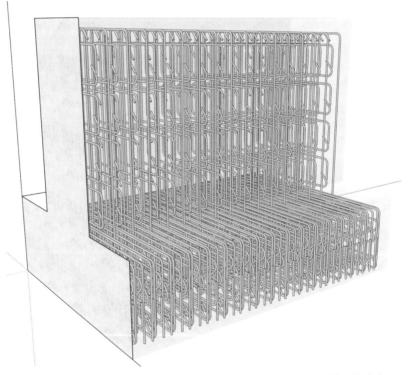

Figure 6.24   Example of a tight pairing of reinforcement bars in a bridge cap. Proper pairing of reinforcement bars is highly important since it will aid you when it is time to use the reinforcement bars in later activities—either estimating or construction.

1. The first step in this workflow procedure is to review the construction or shop drawings and understand how the individual reinforcement bars interact with each other in the pair. For this example, pair one #7 closed stirrup with two #7 ties. The two #7 ties are opposite to each other and lay flat on the closed stirrup (**Fig. 6.25**).

Figure 6.25   Procedure of reinforcement bar pairing.

2. Select the **Move** tool and, while holding down the **Option** button, click on the center of the tie reinforcement and make an offset copy to the side of the original (**Fig. 6.26**).

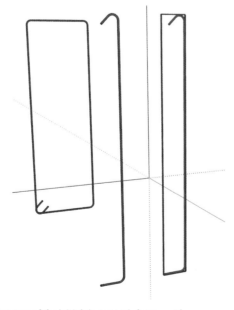

Figure 6.26   Making an offset copy of the initial tie-type reinforcement bar.

3. With the **Scale** tool activated, click on the offset copy and select the middle green square as basis for our modification. Mirror the copied stirrup, by left-clicking on the middle green square and perform a shift in the opposite way until the scale factor in the **Measurements Input Box** reads −1.0 (**Fig. 6.27**).

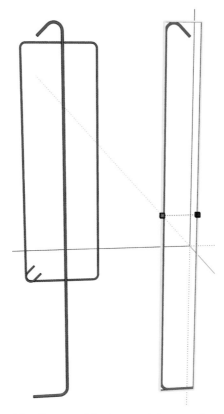

Figure 6.27   Mirror operation of the offset copy of the tie-type reinforcement bar.

Modeling Tip

As discussed previously, SketchUp does not have a standard mirror tool as other CAD systems do but relies on the **Scale** tool to mirror objects about an axis, see Chapter 1-Introduction to SketchUp for more information on the scaling and mirroring commands.

4. For the purpose of this example, the two ties are located 1-ft horizontal offset from the centerline of the main stirrup. In order to accomplish this move, utilize the **Tape Measure** tool in order to add assist lines. The reinforcement bars are already converted into components, and therefore, use the centerline of each of the reinforcement bars as guides—review the first portion of this chapter on how to properly create reinforcement bars (**Fig. 6.28**). Use the main stirrup as a base line for the two tie reinforcement bars.

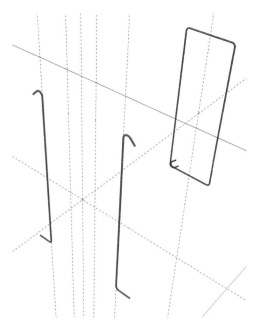

Figure 6.28   Locating the centerlines of the reinforcement bar with the help of the **Tape Measure** tool and the assist lines.

5. After the two tie bars are in the correct location, proceed toward moving them to the center of the main stirrup with the help of the **Move** tool. This will be accomplished by utilizing the assist lines that you created in step 4 (**Fig. 6.29**).

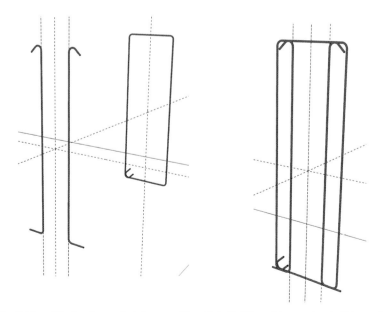

Figure 6.29   Pairing of the two types of reinforcement bars. Note that the pairing overlaps both types of bar.

6. All three reinforcement bars are located on top of each other. The next step is to select the two tie bars and offset them 7/8 in. (one #7 bar diameter) from the center of the main stirrup. Select the two tie bars, and utilizing the **Move** tool, be sure to click on the corresponding arrow key in order to snap the movement to the correct axis; for this example click the right arrow key in order to snap the movement to the **Green** axis (**Fig. 6.30**).

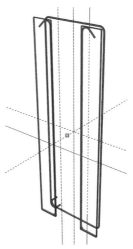

Figure 6.30   Offsetting the tie-type bars to the correct spatial location, using the reinforcement bar diameter as guide for the amount of offset required from the center of the stirrup bar.

7. The next step is to create a component. Select the two stirrups, without selecting the assist lines, and activate the context menu by clicking on the right mouse button. From the context menu select the **Make Component…** option (**Fig. 6.31**).

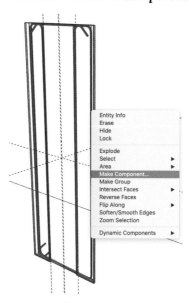

Figure 6.31   Activating the context menu in order to start the component creation procedure for the newly paired reinforcement bars.

8. At the component inspection window, give the proper name and description to the stirrup group and also set the local axis. It is recommended to name the stirrup group with all the individual tag names from which the group is composed, for example, **7C1+7C1+7C2** (**Fig. 6.32**). The reason for this type of naming convention is that it gives you the ability to see all the different components in the rebar group with a single click.

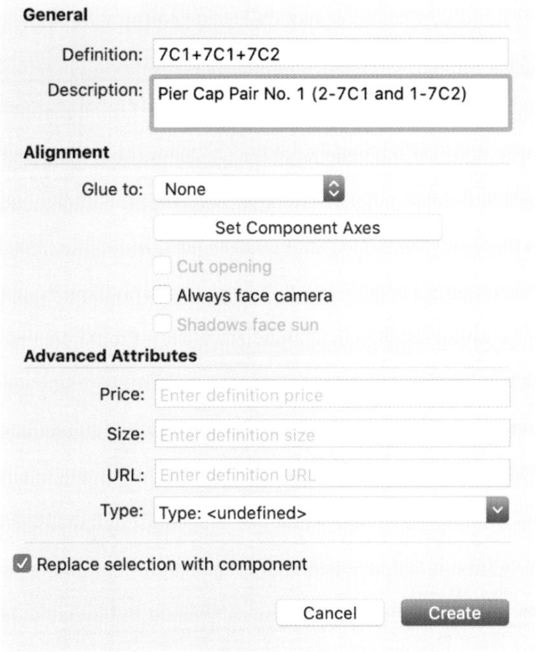

General

Definition: 7C1+7C1+7C2

Description: Pier Cap Pair No. 1 (2-7C1 and 1-7C2)

Alignment

Glue to: None

Set Component Axes

☐ Cut opening
☐ Always face camera
☐ Shadows face sun

Advanced Attributes

Price: Enter definition price

Size: Enter definition size

URL: Enter definition URL

Type: Type: <undefined>

☑ Replace selection with component

Cancel   Create

Figure 6.32   Example of a Description and Definition names for the newly paired reinforcement bars.

9. The next step is to position the local axes of the stirrup group. For this purpose use the assist lines you created in step 5 of this exercise, as location placeholders. From the component inspection window, click on the **Set Component Axis** button (**Fig. 6.33**). The mouse cursor will change from an arrow to the axis symbol. Set the axis cursor in the center of the assist lines and press the left mouse button three times in order to set the x, y, and z axis, respectively (**Fig. 6.33**).

Figure 6.33   Locating the local axes of the paired reinforcement bars. The local axes should be located in the center of the reinforcement group in order to aid the user for easier placement in a three-dimensional object.

10. After the local axes is set for the stirrup group, click on the **Create** button, and the component will be created (**Fig. 6.34**).

Figure 6.34   Converting the paired reinforcement object into a component.

> **Modeling Tip**
>
> The word *"group"* is loosely used in order to denote that it is a pair or group of reinforcement bars and not a function of SketchUp. In Section Three of the exercise, you created a component which is assembled from one stirrup-type bend and two tie-type bend reinforcement bars.

Modeling a Steel-reinforced Concrete Pier

The idea behind this last exercise is to use what you have learned in this chapter till now and add the last knowledge tool needed in order to build a three-dimensional representation of a steel-reinforced pier. For this example, you can either use what you created in the last exercise, with the addition of a new reinforcement pair, or create two new reinforcement pairs; the main purpose is to practice what you did in the last section and at the same time have fun with it!

1. The first step in this exercise is to find the cross-sectional center of the pier. Select the **Tape Measure** tool and left-click on one side of the pier. Utilizing SketchUp's inference midpoint, find the first center. Repeat the step for the second midpoint and the second center. After the center of the Pier is located, place a longitudinal assist line, which will span the entire length of the pier (**Fig. 6.35**).

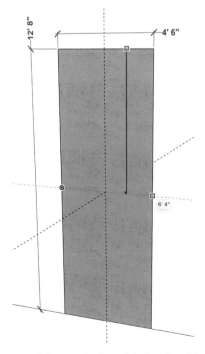

Figure 6.35   **The overall size geometry of the example pier and the location of the pier centerline utilizing the Tape Measure tool and the assist lines.**

2. The next step is to pair the required stirrups. Recall what you did in **Section 3—Pairing of Reinforcement Bars** and repeat the process in this step. You need to pair the following stirrups: **(2)7C1** with **(1)7C2** and **(1)7C2** with **(10)7C3** (**Fig. 6.36**).

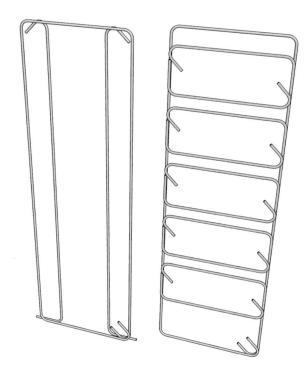

Figure 6.36   Example of two different types of reinforcement pairs that are used in the example.

3. From the assist lines created in Step 1 and utilizing the **Tape Measure** tool, offset another assist line toward the interior of the pier, with a total distance of 4-7/8 in. this will be the starting point of your first stirrup pair.

Modeling Tip

When offsetting the assist line in step 4, care should be taken to account for the clear cover requirements and the thickness of the reinforcement. Do not forget that the centerline of the paired stirrups is located in the center of the pair; therefore, the total offset distance should be 4 in. (clear cover) + 7/8 in. (one bar diameter) = 7/8 in. (**Fig. 6.37**).

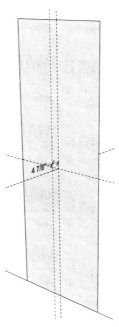

Figure 6.37   Offset distance from the face of the pier structure. The assist line locates the center of the first reinforcement pair.

4. Place the first stirrup pair ((**2**)**7C1** and (**1**)**7C2**) at the center of the longitudinal assist line that was created in Step 1 (**Fig. 6.38**). Note: there will be a total of 20 stirrup pairs in the first section of the beam, spaced at 6 in. on-center.

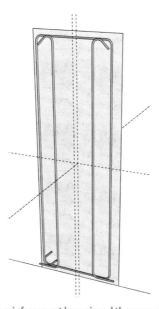

Figure 6.38   Representation of the reinforcement bar pair and the general location in the pier cap.

5. Select the **Move** tool and while holding down the **Option** button, click on the center of the stirrup pair and make an offset copy with a distance of 6 in. from the initial stirrup pair that was placed in step 5. Type the 6-in. distance in the **Measurements Input Box** and press **Enter** (**Fig. 6.39**).

Figure 6.39   Offset copy of the first reinforcement pair and preparing the pair for array copying.

6. Without clicking anything else, type **19X** in the **Measurements Input Box** and press **Enter** (**Fig. 6.40**). SketchUp will automatically add an additional 18 stirrup pairs. This is the same procedure that was used for the steel piles in **Chapter 5— Modeling of Substructure Components**.

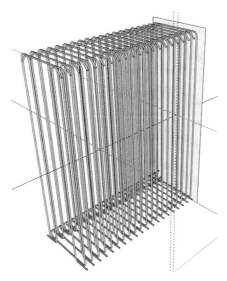

Figure 6.40   Array copying of multiple reinforcement pairs by entering the total number of pairs required.

Modeling Tip

When you use the **NumberX** command in the **Measurements Input Box**, SketchUp will create the same number of copies as is specified by the numeral that is recorded in front of the **X**. Do not forget to count the starting copy for your total number of copies. In this case you need an additional 18 stirrup pairs, but you will have to write **19X** in order to account for the starting stirrup pair from which you will make the additional copies.

7. The next step is to record the new starting point for your next set of stirrup group (**(1)7C2** and **(10)7C3**). In order to accomplish this task, first you need to create an assist line by utilizing the **Tape Measure** tool at the center of the last stirrup group that was created in step 6 (**Fig. 6.41**).

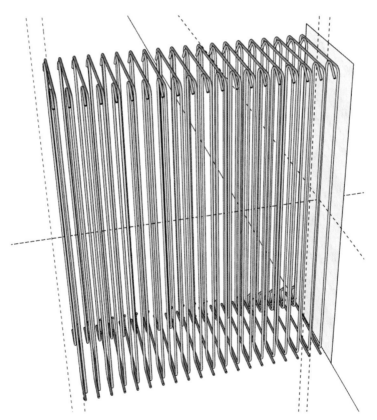

Figure 6.41   Locating the starting point for the second type of the reinforcement pair.

From the newly created assist line, and by utilizing the **Tape Measure** tool again, offset another assist line 6 in. away—this will represent the starting point for your new set of stirrup groups (**Fig. 6.42**).

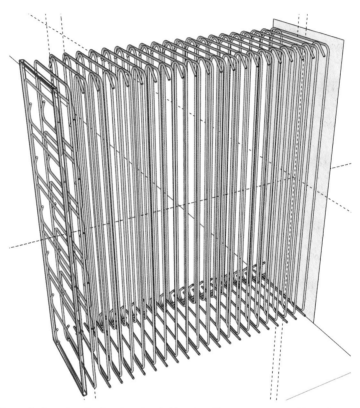

Figure 6.42   Two assist lines placed as location place holders. An offset copy was created from the second type of reinforcement pair in order to prepare it for array copying.

8. Place the initial set of the second group of stirrups at the intersection created by the assist lines in step 7 and repeat the process outlined in steps 5 and 6 (**Fig. 6.43**). (A total of 20 stirrup groups, spaced at 6 in. on-center are needed in the second portion of the pier.) Congratulations, you have completed your first steel reinforced cap.

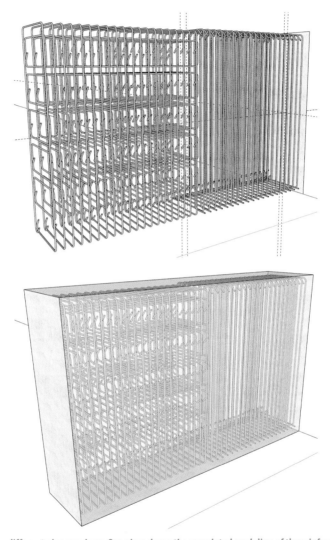

Figure 6.43   **Two different pier cap views. One view shows the completed modeling of the reinforcement pairs, and the second view shows a completed section of the pier cap.**

A couple of closing notes: you can toggle the visibility (eye icon) **OFF** and **ON** from the **Layers** inspector window (**Tags** in SketchUp version 2020) in order to hide certain stirrups for better visualization of the cap during the modeling process.

I encourage you to continue with the exercise by applying what you learned in this chapter. Model the main longitudinal bars!

Modeling Tips for Reinforcement Drawings

The following tips will help you to streamline your modeling experience and minimize possible error that can cost you time:

1. **Analyze the contract and shop drawings:** Spend some time analyzing the structure and the steel-reinforced skeleton underneath. Take a good look at the cross-sections and determine how the individual reinforcement bars come together in the structure you plan to model in three-dimensions.

2. **Use the same naming convention:** When creating different rebar components use the same naming convention as what is shown in the shop or construction/contract drawings. Utilizing the same naming convention will help you when you find a specific conflict or you are trying to explain a situation to field personnel. Since the reinforcement bars are delivered to the job site with name tags based on the shop drawings, it will be very easy to pinpoint the correct bars and make shop or field modifications.

3. **Use actual versus typical rebar sizes:** Keep all the rebar diameter sizes in the three-dimensional model according to what is specified in the shop or contract drawings. This concept requires some premodeling planning, but it will save you a great deal of rework in the long run. The reason for adding extra work at the beginning of the modeling stage is the fact that from the early stages until the end of your project, the three-dimensional model will change the reason why it was created in the first place. It might have been created for elevation and alignment check purposes, but as the project evolves there will be a need to perform clash detection between rebar and post-tensioning ducts or rebar and formwork components. Creating objects in their entirety is a very good work habit to have, but ultimately it is up to the individual modeler on how to proceed forward.

4. **Account for tolerances:** As with everything in the construction industry, there are tolerances in the overall placement of the reinforcement, and the amount sometimes varies between construction authorities. Clear cover should be accounted for between the outer edge of the reinforcement bar and the outer edge of the concrete structure. These two points are specifically important if you are creating a model for clash detection purposes.

5. **Group rebar components:** Very often, several reinforcement bars are grouped together in order to create a unit. This is particularly true for open-type stirrups used in steel-reinforced pier caps. Review the shop or contract drawing on how these groups need to be assembled; follow the work flow outlined in **Section 3** in this chapter.

6. **Complete a final check on the structure:** Account that all the reinforcement bars shown in the contract or shop drawings were used in the three-dimensional model. The **Entity Info** inspector can help you to account for the number of reinforcement bars per type. It's very easy to overlook a section especially on a large and complex structure.

As was mentioned earlier in this section, all of the previously mentioned tips will save you a great deal of headache during the modeling phase. Think about it this way, organization is the key to your productivity because the more you get organized at the begining of the modeling creation stage, the faster you can create your model, which means that you can jump to the next one.

CHAPTER 7

Modeling of Various Bridge Components and Accessories

The idea behind this chapter is to explore different types of modeling procedures for structures and accessories that are most commonly found on bridges and other types of heavy civil projects. Although some of the objects covered in this chapter look very simple to create, they were actually chosen by design to show how a clever use of basic tools can be a substitute for extensions and how proper planning prior to modeling can increase productivity.

Post-tensioning Components

Post-tensioning components can be found in a wide variety of structures, from pier caps to columns, decks, steel supports, and the like. They are primarily used to provide tensile resistance for structures by means of compression. The following section mainly concentrates on three workflows. The first two workflows cover how to create a complex three-dimensional component, as is the case with the anchorage, and how to create spiral-type reinforcement, as is the case with the bursting steel (**Fig. 7.1**).

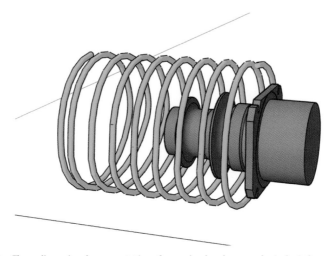

Figure 7.1   Three-dimensional representation of an anchor head, cap, and spiral reinforcement (busting steel).

The third workflow in the series concentrates on the post-tensioning duct layout. Furthermore, it uses the components that were created in the first two workflows to create a complete three-dimensional model of a post-tensioned pier.

Modeling of Anchorage

The following workflow example helps create a plan and models a complex post-tensioning accessory, as is the case for the anchorage. There are a couple of things to keep in mind when modeling the anchorage component of the post-tensioning system for clash detection purposes:

- Unless you are modeling the anchorage to show specific interworking details of the post-tensioning system, you should only be interested in the accuracy of the outside geometry, especially when you are creating a model for clash detection purposes.
- Bundle the cap and heat shrink sleeve together with the anchorage in the same component.
- No need to model the grout/inspection port on the anchorage: it represents a very small detail that will not make a visual or geometric difference if it is not present.

– Try to remove as much of the internal geometry of the anchorage model in order to make the component smaller in size by eliminating edges that will not be visible. Making the component smaller in size will also help with the overall demand on your computer and the response time of SketchUp during your modeling work.

For this example we use a standard type of post-tensioning anchorage. You can use any type of anchorage for this example as long as you have all the necessary views as shown in **Fig. 7.2**. In addition, when importing the two-dimensional drawing, follow the steps outlined in **Chapter 3—Importing Files, AutoCAD, and SketchUp**.

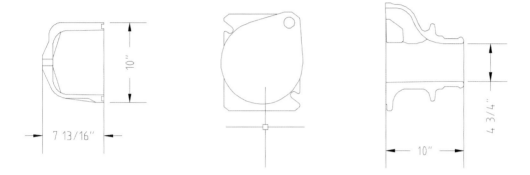

Figure 7.2   Views of different components in a post-tensioning system.

1. The first step is to draw a large rectangular surface around all the parts of the anchorage by utilizing the **Rectangle** tool (**Fig. 7.3**). The size of the rectangle surface does not have to be very accurate as long as it engulfs all the parts of the anchorage.

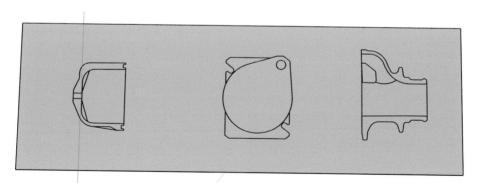

Figure 7.3   Rectangle surface engulfing all of the parts of the anchorage system.

2. The next step is to trace the geometry of the anchorage to the surface. Tracing of the geometry should be straightforward and can be done by utilizing the **Line**, **Arc**, and **2 Point Arc** tools. Only trace the bottom half, outside geometry of the anchorage, and stop the tracing in front of the trumplate, you will create this part later in the example since it has a square geometry, unlike the circular type of the anchorage (**Fig. 7.4**).

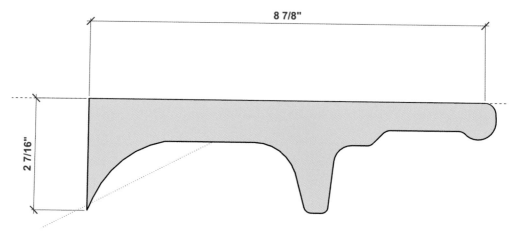

Figure 7.4   Representation of the cross-sectional face of the anchor. At this point in the modeling steps the anchor is in two dimensions.

3. Repeat the outline procedure in step 2 for the cap; only trace the bottom portion (**Fig. 7.5**).

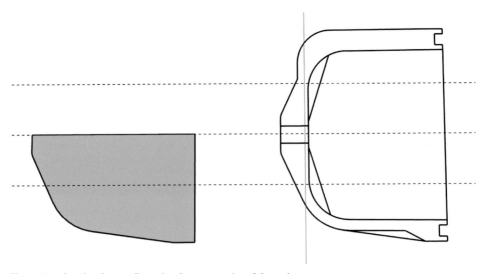

Figure 7.5   Creating the two-dimensional representation of the anchor cap.

4. Utilizing the **Circle** tool, create a circle at the back portion of the traced parts (**Fig. 7.6**). Utilize the correct corresponding arrow keys on your keyboard to lock the axis, prior to selecting the center of the circle.

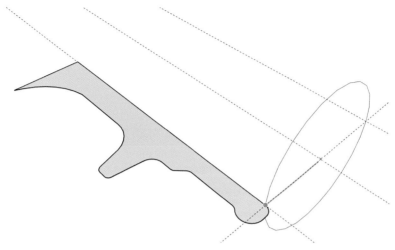

Figure 7.6   Create a circle with the correct diameter (mimic the anchor diameter) in order to act as a path for the Follow Me tool.

Modeling Tip

SketchUp allows you to lock the axis of rotation by clicking on the corresponding arrow keys on the keyboard prior to selecting the point of rotation. The left arrow key will lock on the **Green** axis, the right arrow key will lock on the **Red** axis, and the up arrow key will lock on the **Blue** axis.

5. Select the **Follow Me** tool and left-click on the bottom half of the anchorage. Utilizing the created circle as a guide, extrude the bottom half around the circle; with this you will create the anchorage (**Fig. 7.7**).

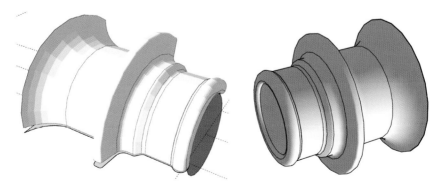

Figure 7.7   Rotating the two-dimensional cross-section around the circle path will create the three-dimensional representation of the anchorage.

6. Repeat the same process as outlines in steps 4 and 5 for the cap (**Fig. 7.8**).

Figure 7.8 Three-dimensional representation of the anchor cap. The anchor cap was modeled utilizing the **Follow Me** tool and a circular guide to act as a path.

7. From the **Styles** toolbar, select the **X-Ray** view or perform a regular visual inspection in order to check that internal lines or geometry are not present in the three-dimensional objects (**Fig. 7.9**). If internal geometry is present, delete it while you are in the **X-Ray** mode. Deselect the **X-Ray** mode when done and move to step 8.

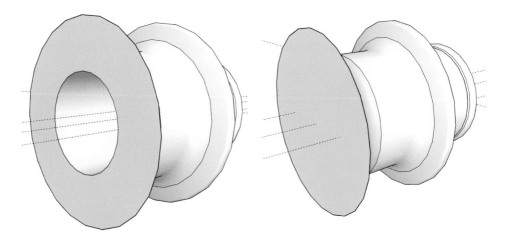

Figure 7.9 Removing unnecessary geometry from the three-dimensional representation of the anchor.

8. Trace the outline of the trumplate by utilizing the **Line**, **Arc**, and **2 Point Arc** tools from the two-dimensional front view. You can also use the **Rectangle** tool and subtract what you do not need from the surface (**Fig. 7.10**). Utilizing the **Push/Pull** tool extrude the trumplate to the correct thickness as shown in the two-dimensional elevation view.

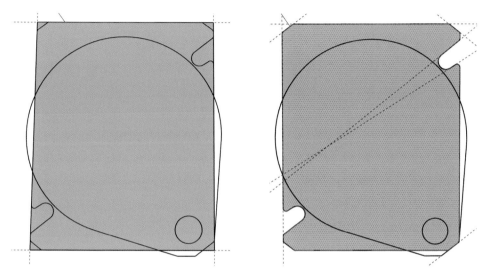

Figure 7.10   Tracing the trumplate and preparing it for extrusion to the correct thickness.

9. The last step is to put the assembly together. Select the individual three-dimensional models, and utilizing the center point of each, assemble them as shown in **Fig. 7.11**. When the assembly is complete proceed toward the last steps to make them into a single component.

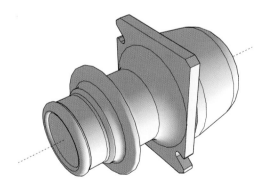

Figure 7.11   Three-dimensional representation of a fully assembled anchor head.

10. Select all four components (cap, trumplate, anchorage, and shrink sleeve) and activate the context menu by clicking on the right mouse button. From the context menu select the **Make Component...** option (**Fig. 7.12**).

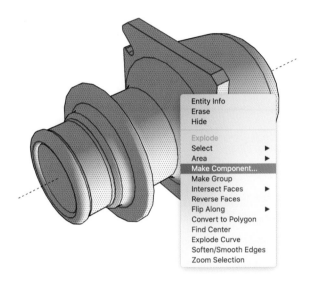

Figure 7.12   Accessing the context menu to convert the anchor object into a component, utilizing the **Make Component...** option.

11. In the component inspection window, type **Anchorage** in the **Name** and **Description** boxes (**Fig. 7.13**).

Figure 7.13   Addition of a unique name and description to the attributes field in the **Create Component** dialog box.

12. The next step is to position the local axes of the anchorage system. From the component inspection window, click on the **Set Component Axis** button (**Fig. 7.14**).

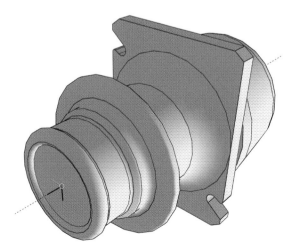

The mouse cursor will change from an arrow to the axis symbol. Set the axis cursor in the center point of the back circle portion of the sleeve by relying on SketchUp's center-point inference. Press the left mouse button three times in order to set the *x*, *y*, and *z* axis, respectively (**Fig. 7.15**).

Figure 7.15   Setting the new location of the local axis. The local axis was located at the connection point with the tendon for ease of installation during future modeling.

13. Congratulations, you have completed the anchorage head (**Fig. 7.16**)!

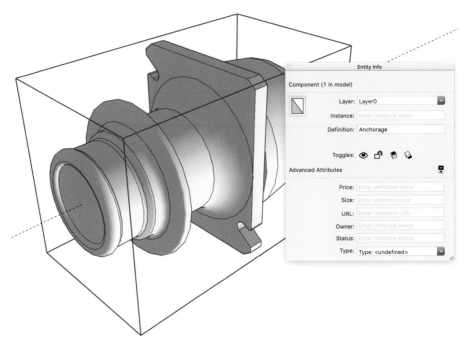

Figure 7.16   Completion of the anchor head component.

Post-tensioning Spiral Reinforcement Bar (Bursting Steel)

The following workflow example goes over the necessary steps to create a spiral type of reinforcement, also known as the bursting steel. The same steps can also be used when creating a spiral reinforcing stirrup for columns.

1. The first step is to review the specification sheet for the spiral reinforcement bar. You are mainly interested in the following properties: diameter of reinforcing bar, number of turns, pitch, and the diameter of the spiral reinforcing bar (from the outside of the bar; **Fig. 7.17**).

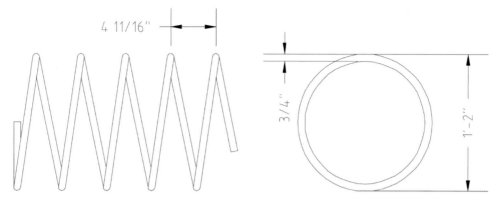

Figure 7.17   Two-dimensional representation of the spiral reinforcement.

2. After you get yourself familiar with the properties, the next step is to create a 13.25-in. diameter arc, utilizing the **Arc** tool (**Fig. 7.18**). Although the overall diameter of the bursting steel is 14 in., you have to take into account the thickness of the bar itself. Recall the procedure for modeling reinforcement bars in **Chapter 6—Modeling of Steel Reinforcement**.

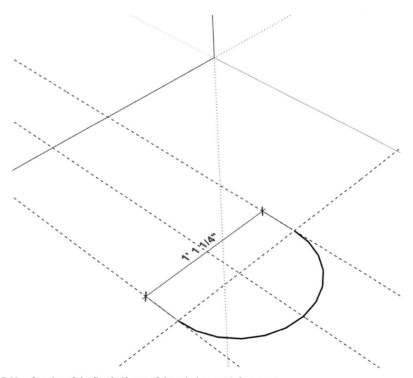

Figure 7.18   Creation of the first half-turn of the spiral-type reinforcement.

3. With the help of the **Tape Measure** tool, create assist lines to mark the pitch for the first turn. Select the **Tape Measure** tool, and offset two assist lines in a vertical direction, each separated by a 2.34-in. distance (each of the two assist lines set the half and full mark of the pitch); use the **Measurements Input Box** to specify the exact distances (**Fig. 7.19**).

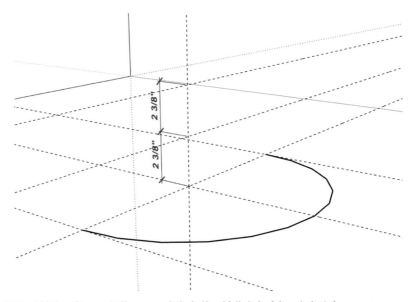

Figure 7.19   Addition of two assist lines to mark the half and full pitch of the spiral reinforcement.

4. The next step is to rotate the arc you created in step 1, which will create the first half of the first spiral turn. Select the arc and then select the **Rotate** tool. Hover over the end point of the arc that is opposite from the vertical assist lines. Press the corresponding arrow key on your keyboard in order to lock the correct rotation axis (**Fig. 7.20**).

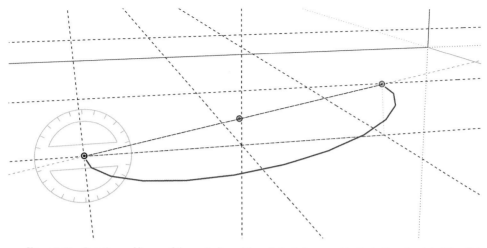

Figure 7.20   Rotation and lineup of the centerline of the spiral reinforcement to the half mark point of the pitch.

Modeling Tip

As mentioned previously, SketchUp allows you to lock the axis of rotation by clicking on the corresponding arrow keys on the keyboard, prior to selecting the point of rotation. The left arrow key will lock on the Green axis, the right arrow key will lock on the **Red** axis, the up arrow key will lock on the **Blue** axis, and the down arrow key locks based on a specific surface.

5. Select the rotated arc, and utilizing the **Move** tool, make an offset copy by holding down the **Option** key on the keyboard. The offset distance is not important at this point (**Fig. 7.21**).

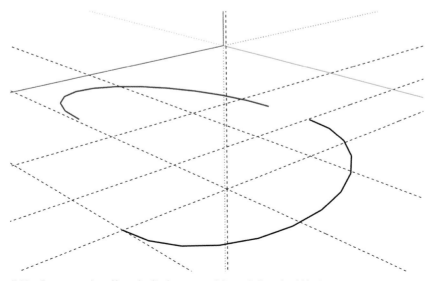

Figure 7.21   Representation of how the final outcome of the copied arc should look.

6. Select offset copy of the arc, and utilizing the **Scale** tool, mirror the offset copy by selecting the diagonal green selection boxes (**Fig. 7.22**).

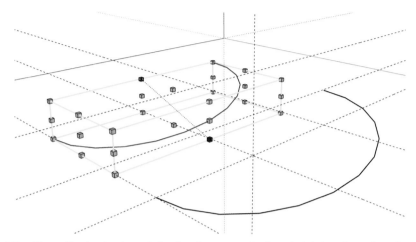

Figure 7.22   Diagonally mirroring the copied arc in order to create the first spiral pitch, utilizing the **Scale** tool.

7. Select the mirrored copy of the arc; utilize the **Move** tool and place it at the end point of the original arc (**Fig. 7.23**).

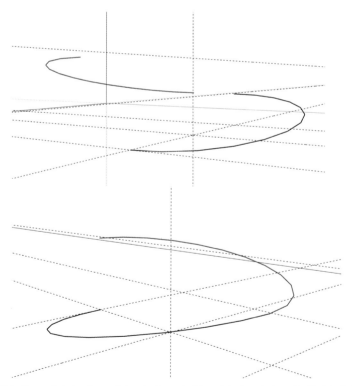

Figure 7.23   Selecting and placing the copied arc to the end of the original created arc. The first piral pitch is created.

8. Select the newly created first pitch of the bursting steel, and utilizing the **Move** tool, make an offset copy by clicking on the **Option** key (**Fig. 7.24**). Without deselecting the copied pitch, in the **Measurements Input Box** type 5× and press Enter. SketchUp will automatically generate four additional pitch turns.

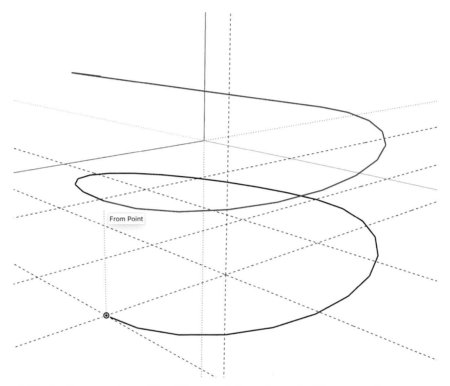

Figure 7.24   Creating a second copy of the pitch representation and preparing it for array copying.

9. The next step is to create the diameter of the reinforcing bar. Select the **Circle** tool and click on the lowermost point of the spiral; set the diameter to 3/4 in. by entering it in the **Measurements Input Box**. If needed, lock the correct axis by clicking the corresponding arrow keys on the keyboard prior to selecting the center point of the circle; see the **Modeling Tip** for more information. The created circle has to be perpendicular to the spiral line (**Fig. 7.25**).

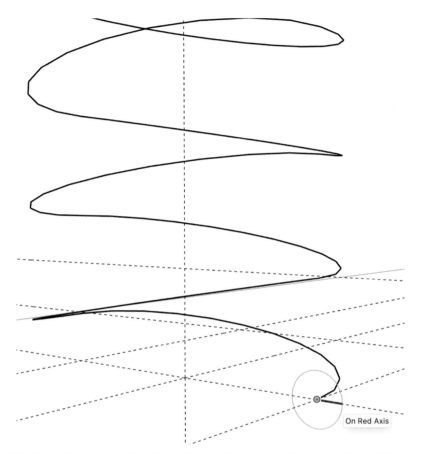

On Red Axis

Figure 7.25 The skeleton portion of the spiral reinforcement is shown complete. A perpendicular circle was created to denote the thickness of the reinforcing bar.

Modeling Tip

As mentioned previously, the default number of sides for the **Circle** tool is set at 24 in the SketchUp application. Adjust the number of sides to a lower value, depending on the amount of post-tensional ducts that have to be modeled in order to save on computer resources and increase the response time of SketchUp. See **Chapter 1** for more detail on how to change the number of sides.

10. Select the **Follow Me** tool and left-click on the created circular surface. Utilizing the created spiral edge, extrude the circle until it reaches the endmost point (**Fig. 7.26**).

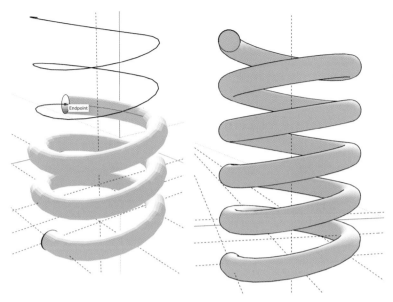

Figure 7.26   Representation of the spiral reinforcement during and after the use of the **Follow Me** tool.

11. The final steps will be to create a usable component. Select the entire length of the spiral reinforcement, without selecting the assist lines that create the centerline and activate the context menu by clicking on the right mouse button. From the context menu select the **Make Component...** option (**Fig. 7.27**).

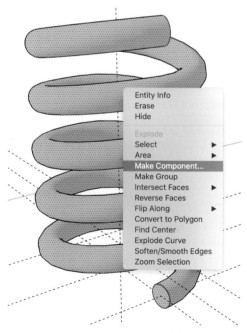

Figure 7.27   Accessing the context menu in order to convert the model into a component for future use, using the **Make Component...** option.

12. At the component inspection window, type **Type I Spiral Bar** in the **Name** and **Description** boxes (**Fig. 7.28**).

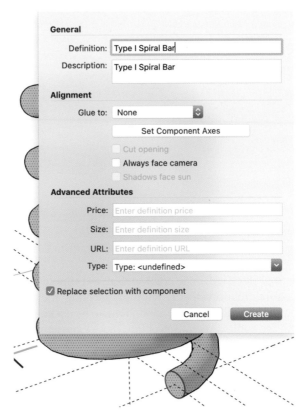

Figure 7.28   Adding the proper information to the **Definition** and **Description** fields in the Make **Component** dialog box.

13. The next step is to position the local axis of the spiral reinforcement bar. For this purpose you will use the assist lines that you created in step 2 of the exercise, as location placeholders. From the **Make Components...** inspection window, click on the **Set Component Axis** button (**Fig. 7.29**). The mouse cursor will change from an arrow to the axis symbol. Set the axis cursor in the center of the assist lines and press the left mouse button three times in order to set the x, y, and z axis, respectively (**Fig. 7.29**).

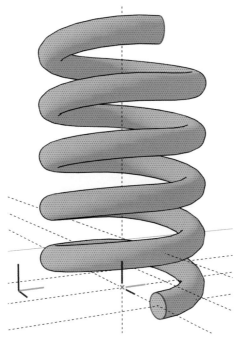

Figure 7.29   The new location of the local axis for the spiral reinforcement. The new location will make it easier to place the spiral reinforcement in the correct place on the anchorage.

14.   After the local axis is set on the spiral reinforcement bar, click on the **Create** button, and the component will be created (**Fig. 7.30**).

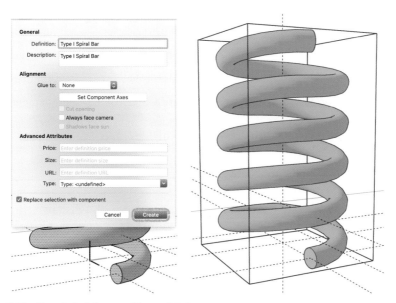

Figure 7.30   The spiral reinforcement is completed.

Layout of Post-tensioning Ducts and Creation of the Entire Unit

The duct for the wire strand represents the last key element in the post-tensioning system. With its corrugated shape, which provides bonding properties for the concrete and grout, it snakes throughout the pier cap at specific points of transition, which are determined by structural analysis. Because of these structural specifics, accurate modeling is a must when dealing with post-tensioning ducts and the post-tensioning system in general (**Fig. 7.31**).

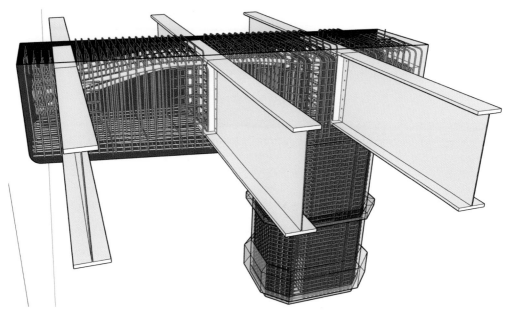

Figure 7.31 Three-dimensional representation of a complex pier with post-tensioning ducts located in the upper portion. The main pier reinforcement was hidden for visual purposes.

For this workflow example we use the two-dimensional representation of the cap shown in **Fig. 7.32**.

From experience, the seven key points required to be successful in your modeling endeavor are as follows:

1. Review the post-tensioning system, prior to starting. Fully understand how the ducts are moving inside the objects.

2. Always redraw the tendons yourself utilizing the distances in the contract or shop drawings. Never import contract drawings for this task, since most often they can be improperly scaled when printed or never revised if a change was made: by the designer.

3. No need to represent the corrugation on the duct, as it is too time-consuming and it will drain your computer's resources.

4. When drawing the duct, always use the radius to the outside of the corrugation; this is especially important if you are creating a model for clash detection.

5. Always use the proper radius arcs for the transition points. Transitions points are shown at sharp angles on contract drawings, rather than arcs. This is highly misleading, and it represents the basis for a lot of conflicts between the duct and the surrounding reinforcement bars.

6. For each post-tensioning tendon unit, create a corresponding layer and a component. This is very useful especially when the next task on your list is to model the reinforcement bars.

7. Always use the same naming convention as specified in the contract plans or the shop drawings. This will save you a lot of headaches when you have to present your findings.

Modeling Work

1. Following the previously mentioned key points, start your first step with a review of the imported two-dimensional contract drawings. Use the drawing for review purposes only of the prestressed tendons. The pier cap has a total of four tendons with two different layouts: two tendons start from the bottom of the pier cap, and another two start from the top. In addition, each tendon type has a total of two transition points (**Fig. 7.32**).

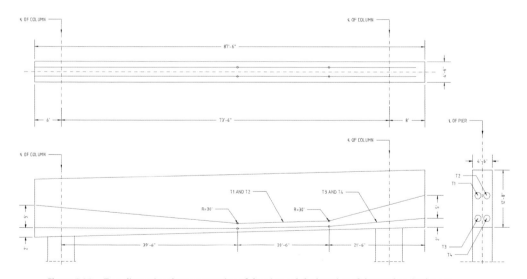

Figure 7.32   Two-dimensional representation of the pier and the location of the tendons in the cap.

2. The second step is to model one longitudinal face of the pier cap. For this purpose you can trace the pier cap directly from an imported drawing or model it from scratch, following the dimensions in **Fig. 7.32**. Do not forget to check dimensions as you trace. Also notice that the pier cap is on an incline, going from left to right (**Fig. 7.33**).

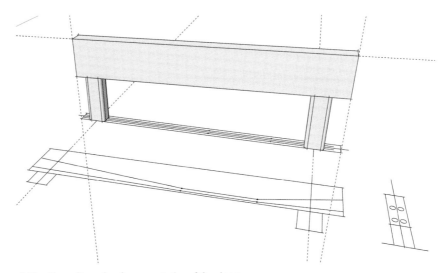

Figure 7.33   Three-dimensional representation of the pier cap.

3. The next step is to trace the path of the tendon utilizing assist lines. Work on one tendon type at a time. Select the **Tape Measure** tool, and using the vertical side of the cap as a guide, offset two assist lines at the location of the transition points. In addition, add two assist lines in order to represent the start and end point of the tendon, following the dimensions in **Fig. 7.32.** The final step is to use the **Line** tool and fill between the assist lines. The reason for the assist line is the possibility that gaps are present if you are working from an imported AutoCAD file as is the case in **Fig. 7.34.** Always remember to use AutoCAD files that are created by you based on the shop/contract drawings as mentioned at the beginning of this section in point 2 of the seven key points.

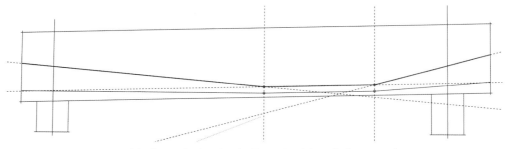

Figure 7.34   Tracing of the first tendon by using the **Line** tool and the assist lines as markers.

4. In step 3 you created the rough outline of the first type of tendon. The next step is to create the specified transition bend radius. Per the contract drawings, the bend radius at the transition points is set at 30 ft. Select the **Tape Measure** tool and offset the angled assist line, created in step 3, parallel for a distance of 30 ft. Repeat the same steps for the other side of the cap. The intersection between the newly offset lines represents the center of the transition bend radius (**Fig. 7.35**).

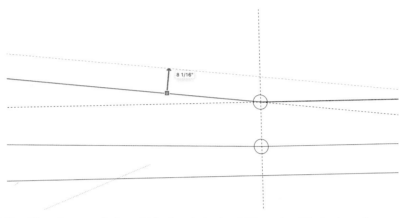

Figure 7.35   Offset of a perpendicular assist line in order to pinpoint the center of the tendon radius.

5. Select the **Circle** tool, and utilizing the intersection of the two assist lines created in step 4, draw a circle with a radius of 30 ft. The intersection between the circle and the tendon represents the curved portion of the tendon geometry (**Fig. 7.36**). Repeat the same step for the other side of the cap. Erase the outside portion of the circle and the extra draw-in lines which are not necessary.

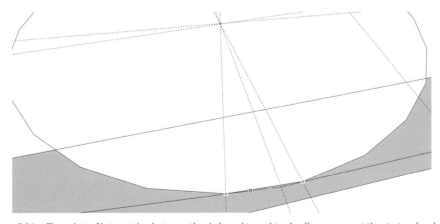

Figure 7.36   The points of intersection between the circle and traced tendon lines represent the start and end point of the tendon curvature (chamfer).

6. The next step is to create the thickness of the tendon. Select the **Circle** tool, and create a circular surface which is perpendicular to the tendon line created in step 5 (**Fig. 7.37**). The circular surface should have an overall diameter of 4-3/4 in.

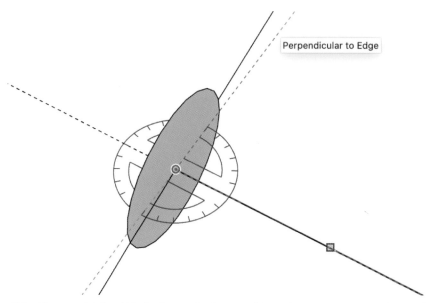

Perpendicular to Edge

Figure 7.37 The created circle, which also denotes the diameter of the tendon, has to be positioned perpendicular to the tendon line. Assist lines can be a great help to achieve this requirement.

7. Select the **Follow Me** tool and click the circular surface created in step 6. Using the tendon line as a guide, extrude the tendon, from one side of the pier cap to the other (**Fig. 7.38**).

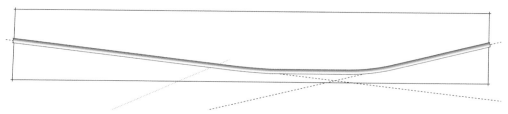

Figure 7.38 Three-dimensional representation the first tendon. The tendon was extruded using the **Follow Me** tool and the tendon line a guide.

8. Select the newly created tendon only (do not select any assist lines), and with a right mouse click, select the **Make Component…** from the context menu. In the **Name** option box write **Tendon 1**, and in the **Description** field write **T1 & T2** (**Fig. 7.39**). You do not have to assign a specific axis location; therefore, click on the **Create** button.

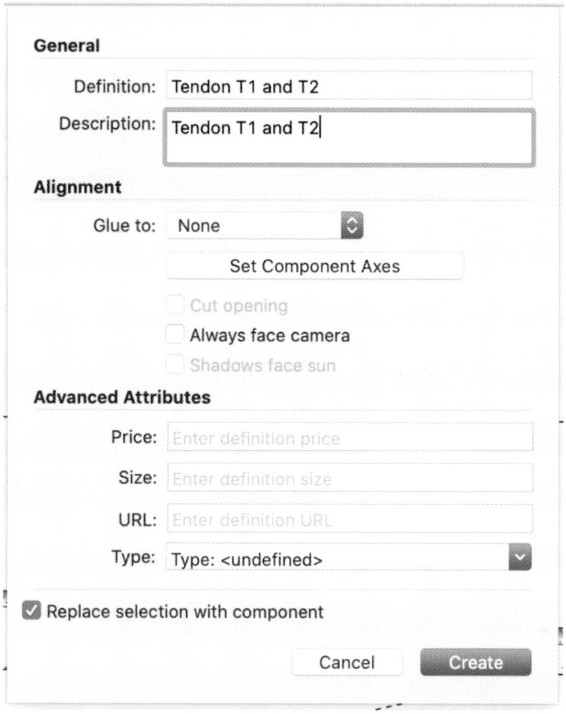

Figure 7.39 Adding information to the **Definition** and **Description** fields in the Make **Component** window.

9. The next step is to add the post-tensioned models of the parts that you created in the last two sections: the anchor head and the spiral reinforcement. You can import them from your own components library, utilizing the **Components** inspection window. Review **Chapter 1—Components and Groups** for the necessary steps to import components.

10. Select the anchored component, and utilizing the **Move** tool, align the center of the anchor head with the center of the tendon. Use the **X-Ray** view mode for easier alignment. Select the anchor head and then click on the **Rotate** tool. Utilizing the arrow key buttons, select the correct plane of rotation (**Fig. 7.40**). For a starting point, click on the intersection between the anchor head and the tendon. For your end point, click on the centerline of the circular surface located at the end of the anchor. Rotate the anchor head until the end point aligns with the centerline of the duct. SketchUp's inference engine will help you identify the end point of the rotation. Repeat the same steps for the other side of the pier cap (**Fig. 7.40**).

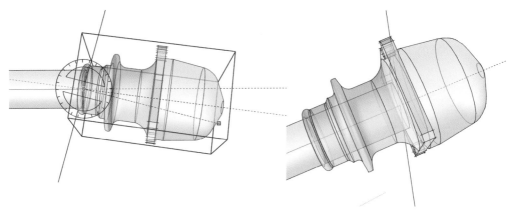

Figure 7.40   Before and after views of the anchor.

11. Select all of the post-tensioning components only (do not select any assist lines), and with a right mouse click, select the **Make Component…** option from the context menu. In the **Name** and **Description** fields write **T1 & T2**. You do not have to assign a specific axis location; therefore, click on the **Create** button to finalize the post-tensioning component.

12. Repeat steps 3 to 11 for tendon lines **T3** and **T4** (**Fig. 7.41**).

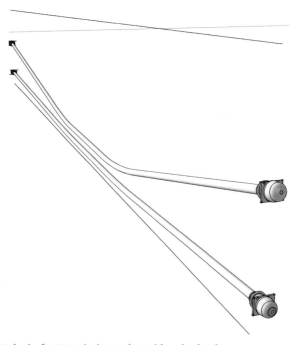

Figure 7.41   Completed pair of post-tensioning tendons with anchor heads.

13. The last step is to place the tendons in the correct location by utilizing the help from the already created assist lines in the three-dimensional model of the pier cap that you created in Step 2 (**Fig. 7.42**).

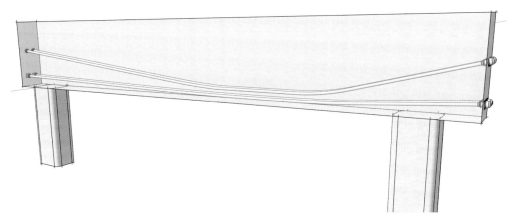

Figure 7.42 Completed three-dimensional representation of the pier cap and post-tensioning tendons.

Step 13 finalizes the workflow example for the creation of a post-tensioning component in a pier cap. For any brave modelers out there, I would encourage you to continue the exercise by adding the pier cap reinforcement bars (**Fig. 7.43**) and applying what you learned in **Chapter 6—Modeling of Steel Reinforcement,** and don't forget to have fun with it!

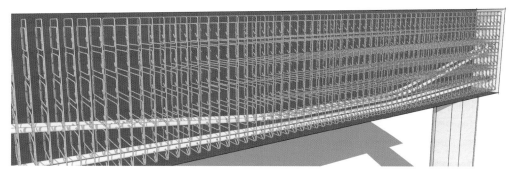

Figure 7.43   There-dimensional representation of the pier cap, post-tensioning tendons, and reinforcement.

Drainage and Fire Suppression Components

The drainage and fire suppression systems represent essential components of heavy civil projects, especially bridge structures. Both systems extend the overall life span, by removing the atmospheric water from the structure and thus slowing down the processes of corrosion, and by preventing fire damage to occur on key structural components Fig. 7.44. shows a three-dimensional representation of a fire suppression riser on a bridge. (**Fig. 7.44**).

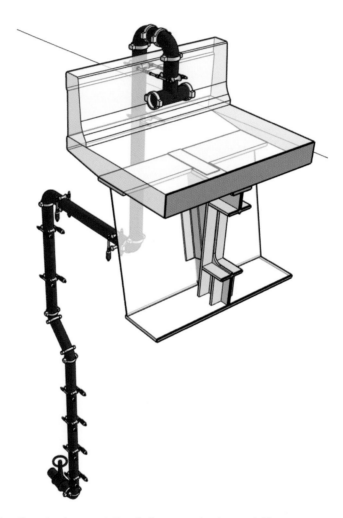

Figure 7.44   Thee-dimensional representation of a fire suppression riser on a bridge.

Because of the important part these two systems play for long-term serviceability of the structure, you will see more and more detailed design on current and future projects to come. The following section concentrates on the proper workflow to model a drainage elbow. Besides a piece of a pipe, the elbow represents one of the most common elements in eiter a drainage or fire suppression systems and therefore was chosen specifically for this workflow. The same steps can be reapplied when modeling elbows with different bend angle configurations or made from different material, as would be the case for fire suppression systems.

As with other chapters in this book, one key rule to remember is the balance of detail versus file size in a given three-dimensional object. The object has to mimic the overall geometric look and size of the component that you are modeling, but you also have to control the overall size of the file, which in turn will govern the responsiveness of your system.

There are a couple of ways to accomplish these goals: one way is to remove the geometry that is not visible, for example, the internal geometry of a pipe or elbow component. You might think this is a futile endeavor, but when you have over one hundred of these components in a model it makes a big difference. Another way is to rely as much as possible on components for objects that will be copied multiple times. As was explained in **Chapter 1**, components will decrease the overall size of the file but at the same time help you if you need to make a substitution in the type of component used because of a revision.

The preceding rule only governs if the drawing is used for layout purposes (to estimate the amount of pipes, couplers, fittings, and other components needed for the entire run of the system). If the three-dimensional model is used for visual purposes to explain a working process of a component, then you will have to add additional true-to-life detail.

For the workflow examples, a typical component from the drainage systems was chosen in order to show the amount of detail you should use in models for layout purposes. Since the fire suppression system and the drainage system share a common thread, and that is they are collection of pipes and elbows, you can use the same workflow when modeling fire suppression components. Following are the instructions for a drainage elbow.

Drainage Elbow

1. The first step is to set the correct layout of the drainage elbow. A 45-degree elbow will be used for this workflow example (**Fig. 7.45**). Select the **Rectangle** tool and draw a 20-in. × 20-in. face.

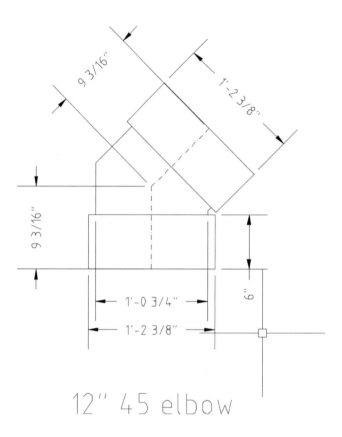

Figure 7.45   **Two-dimensional representation of a 12-in. diameter, 45-degree drainage elbow.**

2. Create the centerline of the elbow by utilizing assist lines with the help of the **Tape Measure** tool. Use perpendicular assist lines to denote the start and end points of the curve in the elbow and the sockets, if needed. Some elbows do not have a straight section but only a curved section—as is the case with the 45-degree elbow in the example.

3. Select the **Line** tool and draw in the centerline of the elbow (**Fig. 7.46**). After the centerline is completed, use the **2 Point Arc** to create the curvature of the elbow (**Fig. 7.46**). The curvature of the elbow has to be tangent to the edge. Use SketchUp's inference engine, by hovering over the edges in order to find the location where the arc will be tangent to the edge. When the correct spot is located, the tangent line will turn magenta (**Fig. 7.46**).

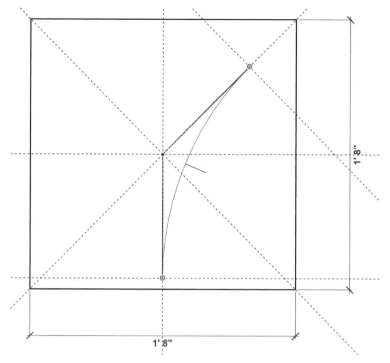

Figure 7.46   Creation of the elbow bend, using the **2 Point Arc** tool and the assist lines.

4. Utilizing the **Circle** tool, create a circle perpendicular to the centerline created in step 3. Click on the corresponding arrow key to lock the correct axis in order to create a perpendicular circle to the line. Although pipe is measured by the I.D. (Inside Diameter), do not forget to use the O.D. (Outside Diameter) dimension when drawing the circle (**Fig. 7.47**).

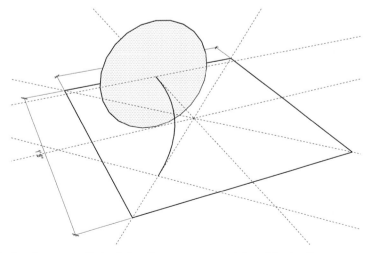

Figure 7.47   Creating a perpendicular circle to the arc, as a representation of the pipe diameter.

5. Click on the **Follow Me** tool, select the created circle, and utilizing the centerline as a guide, extrude the pipe (**Fig. 7.48**).

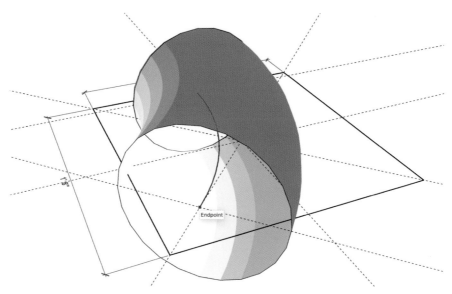

Figure 7.48   The circle was extruded using the **Follow Me** tool and the arc line as a guide.

6. Utilizing the **Circle** tool, create a secondary, larger circle, at each end of the elbow—this is the beginning of the sockets (**Fig. 7.49**).

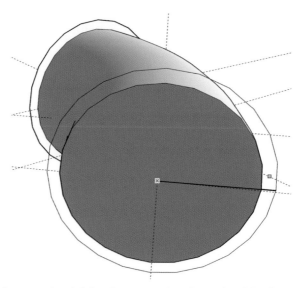

Figure 7.49   Offsetting a secondary circle in order to create the socket portion of the elbow.

Extrude these circles with the **Push/Pull** tool until they reach the beginning of the curvature (**Fig. 7.50**).

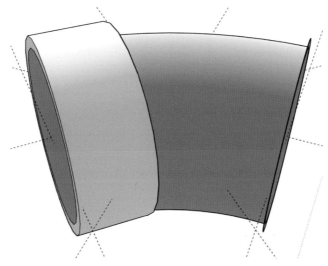

Figure 7.50   Extrusion of the created circle (socket) utilizing the **Push/Pull** tool.

7. Delete the extra lines located in the socket area from the elbow. Click on the **X-Ray** view to help you for this process (**Fig. 7.51**).

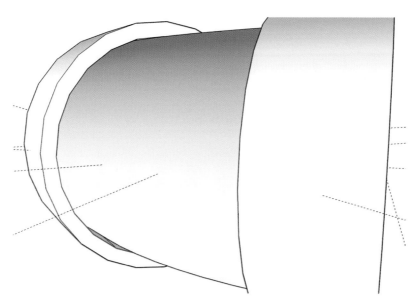

Figure 7.51   Removing all unnecessary lines from the elbow geometry.

8. The last step is to create a component. Select the entire elbow and click on it with the right mouse button to activate the context menu. From the context menu, select **Make Component...** (**Fig. 7.52**). Type in the corresponding information, such as type of elbow (90 degree, 60 degree, 45 degree, and so on), material type, and overall diameter, in the **Definition** and **Description** areas. Click on the **Create** button to complete the process.

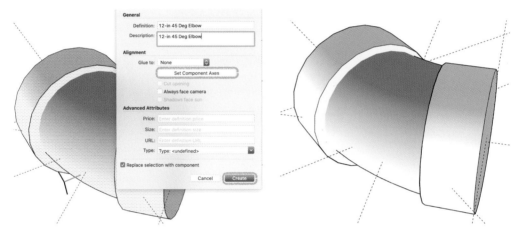

Figure 7.52   Creation of a component the completed drainage elbow.

Modeling Tip

As was mentioned in the introduction portion of this chapter, the removal of the interior lines at the socket area in step 7, which are also not visible under normal circumstances, will help minimize the size of the component, which in turn will minimize the overall size of the models when multiple components are used for layout purposes.

Modeling of Parapet Wall

At first glance, bridge parapet or barrier wall looks like a very simple object to model. The reason why a parapet was chosen as one of the examples in this chapter is the fact that the SketchUp application does not have fillet or chamfer tools, as is the case with other CAD applications. Besides parapets, there are other objects that use fillet and chamfer in their geometry, as is the case for prestressed concrete girders and overhang portions of steel-reinforced decks, just to name a few. The modeling of girders is covered in more detail in **Chapter 8—Modeling of Bridge Decks and Girders** of this book.

One option to consider, as a possible workaround to this problem, is to download and install a third-party extension that will give you the ability to add the chamfer and fillet tool in your toolbar. Although this workaround represents two very viable options, the question is, What to do if you do not want to or you are not able to use third-party extensions?

The following example goes over a detailed workflow on how to draw a 32-in. F-Shape parapet wall, utilizing the standard tools in SketchUp (**Fig. 7.53**).

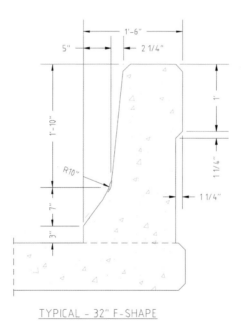

TYPICAL - 32" F-SHAPE

Figure 7.53 Two-dimensional representation of a typical 32-in. F-shape barrier wall.

1. The first step is to draw at least a 20-in. × 40-in. surface by utilizing the **Rectangle** tool (**Fig. 7.54**). The size of the surface will correspond to the overall cross-sectional size of the parapet wall. For this example, the cross-sectional size of the parapet is 18 in. (width) by 32 in. (height); therefore, a 20-in. by 40-in. surface will do fine.

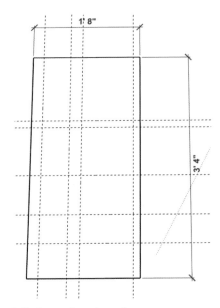

Figure 7.54   Adding assist lines to the base rectangular surface.

The next step is to draw several assist lines utilizing the **Tape Measure** tool. The idea behind the assist line is to draw a rough outline of the barrier wall; this is very helpful since the barrier wall has a lot of angled geometry (**Fig. 7.54**). Also the upper dimensions of the barrier wall intersect the two lines, which cannot be seen from the chamfer.

2. After all the assist lines are drawn, draw in the actual lines over the assist lines utilizing the **Line** tool. Draw-in all the sides of the parapet wall (**Fig. 7.55**).

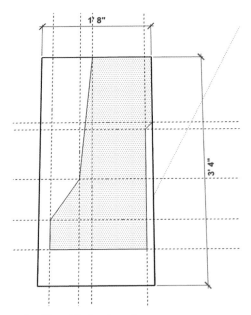

Figure 7.55   Tracing of the rough outline of the barrier wall.

3. For the next step, turn you attention to the front of the barrier, especially the area that has the large fillet (circled geometry). Based on the drawings, the fillet has a radius of 10 in., with the most inward point located 5 in. from the front of the parapet (horizontal measurement) and 10-in. from the bottom of the parapet (vertical measurement). Utilizing the **Tape Measure** tool offset the assist line **1** and **2**, for a total distance of 10″ as shown on **Fig. 7.56**. The point where these two assist lines meet represents the radius of the fillet (**Fig. 7.56**).

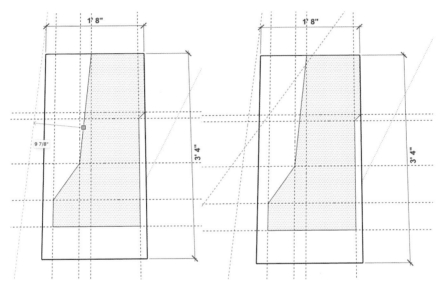

Figure 7.56   (a) Adding assist lines to locate the origin of the curvature (b) The completed assist lines are shown in light blue color and create the origin of the circle that will create the front curvature of the parapet wall.

4. Utilizing the **Circle** tool, draw a circle with a 10-in. radius, utilizing the intersection point of the asset lines from step 3 as the center. The location where the circle is in contact with the barrier line represents the curved geometry. Delete the excess circle and the extra barrier lines beneath the curve geometry (**Fig. 7.57**).

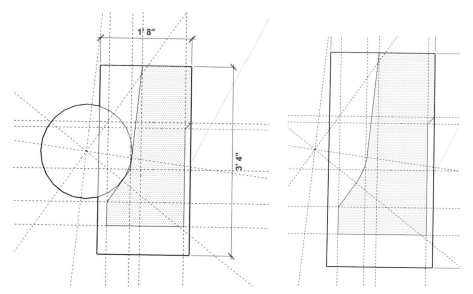

Figure 7.57   Before and after representation of the curvature of the wall. The intersect points between the circle, and the lines represent the start and end point of the arc.

5. The final step is to draw the chamfers on the top and the back side of the parapet, utilizing the assist lines and the **Line** tool (**Fig. 7.58**).

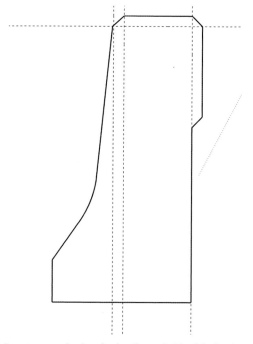

Figure 7.58   Adding assist lines to create the chamfer detail at each side of the barrier wall.

The barrier wall is complete **Fig. 7.59(a)**. I encourage you to continue with the exercise and draw in the reinforcement bars as we practiced in **Chapter 6—Modeling of Steel Reinforcement, Fig. 7.59(b)** shows an example of a "Jersey" type barrier with the coresponding steel reinforcement modeled.

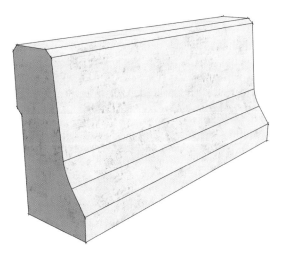

Figure 7.59(a)   Three-dimensional representation of the completed barrier wall.

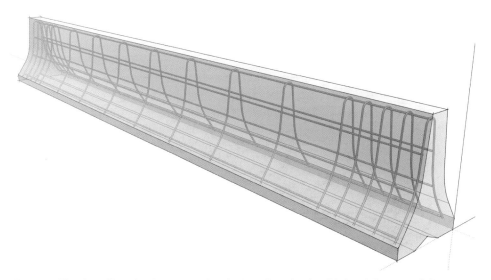

Figure 7.59(b)   Three-dimensional representation of a "Jersey" type barrier with the reinforcement skeleton modeled underneath.

Formwork Components

The main idea behind the formwork section is to showcase a series of steps, which can be equally used during design or construction phases, on how to organize your three-dimensional model for member recognition. Three-dimensional modeling of lumber is straightforward; basically you will use a series of rectangular cuboids to represent different lumber sizes in your model; therefore, this process will not be reviewed in detail. On the contrary, our main objective is to review the organizational portion of the work.

There are a total of two types of formwork used today in construction, a prefabricated type, constructed from steel elements, and a more traditional type, constructed from lumber elements. Regardless of which type of formwork is used for design or construction purposes, the outlined workflow will do the job efficiently to organize your drawings.

Organization Steps

There are several steps that need to be taken into consideration when organizing three-dimensional models of formwork:

- **Use the correct size of lumber or prefabricated component**—Always use the actual size and not the nominal size when creating individual lumber or prefabricated component models. This will make a big difference when performing constructability checks.

- **Create individual layers**—Create individual layers for each lumber size or prefabricated component that you plan on using in your model. This will help in the future when you want to obtain a quantity of a specific lumber size or formwork component.

- **Create individual components**—Similar to the previous step, create individual components for each lumber size or prefabricated component that you plan to use in your model. This will decrease the overall size of your file and will help you if design changes are made to specific lumber sizes or formwork component.

- **Assign colors**—Assign different colors for different types of lumber or prefabricated components. Colors can be assigned based on lumber sizes, structural use (members in compression vs. tension), or geometry (longitudinal vs. transverse).

Organizational Workflow Example

As mentioned at the beginning of this chapter, the overall organization aspect is very important when modeling formwork. For this purpose, this example reviews the overall organizational workflow of a formwork.

1. The first step in the exercise is to review the formwork design. As you can see, the formwork design shown in **Fig. 7.60** comprises of lumber and steel support components.

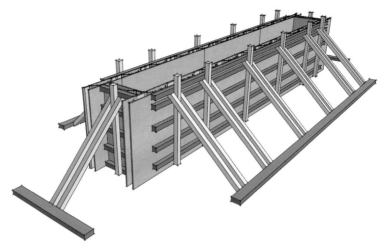

Figure 7.60   Three-dimensional representation of a formwork for a support wall comprised of prefabricated panels and lumber elements.

2. The next step is to review the **Entity Info**, **Components**, and **Layers** (**Tags** in SketchUp version 2020) inspection windows. They can be accessed from the main menu, by left-clicking on **Window** and then selecting **Entity Info** (**Fig. 7.61**).

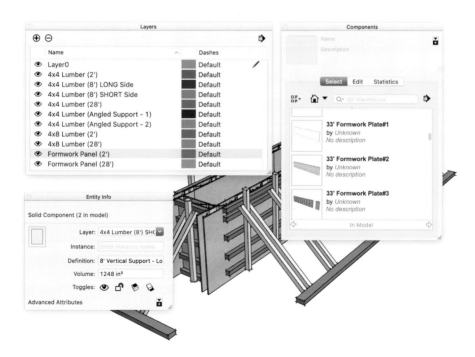

Figure 7.61   Layer (Tags in SketchUp version 2020), Components, and Entity Info inspection windows and the associated data in each of these windows.

3. If you draw your attention to the **Layers** inspection window (**Tags** in SketchUp version 2020) you can see that besides the default layer, **Layer0**, layers were created for each individual lumber size and steel components. In addition, each created layer has a different color block assigned to it (**Fig. 7.62**).

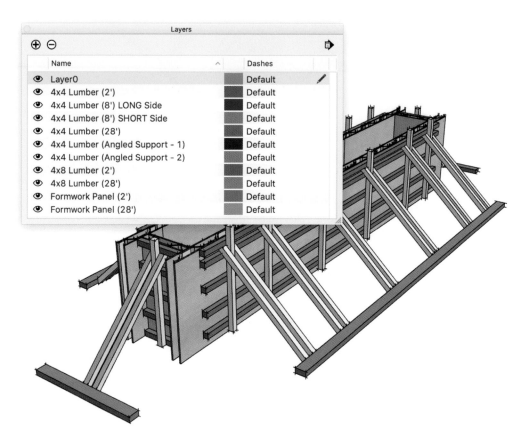

Figure 7.62 List of layers that were created based on the components used in the formwork design.

- Creating individual layers for each lumbar size will allow you to toggle the visibility of the components that are assigned under that particular layer name. This is especially useful when you have more complex formwork, where during construction you will want to "peel off" the outer layer in order to show the internal supports without modifying the drawing (**Fig. 7.63**).

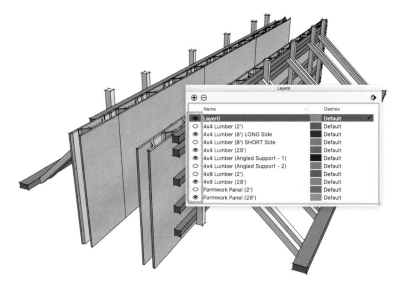

Figure 7.63 The ability to hide and show different layers of the formwork design using the eye icon in the **Layers** inspection window (**Tags** in SketchUp version 2020). The ability to hide certain components, allows you to create work plans which will show the step-by-step instruction on how to assemble the formwork, or hide certain sections during constructability review stages.

- Assigning a different color block to each layer will help you better visualize the different lumber sizes or prefabricated panels and where they are located in the formwork. This can be accomplished by activating the option **Color by Layer,** which can be found under the **Layers** context menu (**Fig. 7.64**).

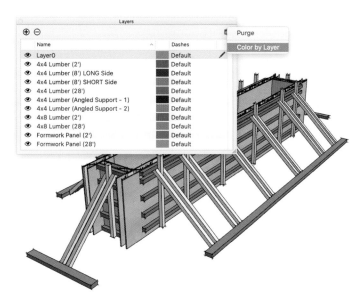

Figure 7.64 Different visual markers are assigned to different layers for ease of visual recognition of structural members in the formwork.

When assigning a color block, keep in mind to assign contrasting colors; this way it will easily focus your attention during the review process (**Fig. 7.65**).

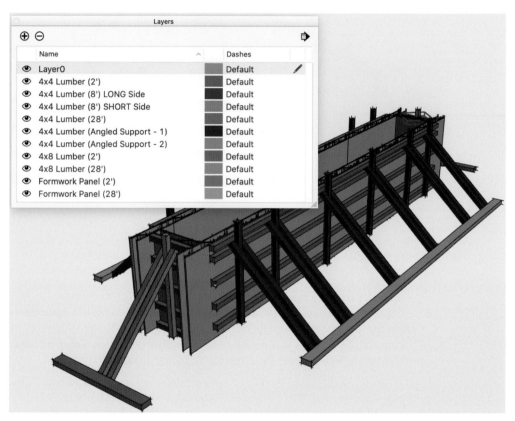

Figure 7.65   Activating the **Color by Layer** option in the **Layers** inspection window (**Tags in SketchUp version 2020**) allows you to differentiate between different components by the color code.

4. As was mentioned in the **Organizational Steps**, besides assigning individual layers, each lumber member was converted into a component. Having each lumber converted into a component gives you three separate advantages:

    • The first advantage is the ability to receive an accurate count of the number of lumber pieces used in the formwork. This can be especially helpful for procurement purposes. Select any lumber piece on the model and draw your attention to the **Entity Info** inspection window. In the upper-left corner of the window, SketchUp will display the number of components used in the model (**Fig. 7.66**).

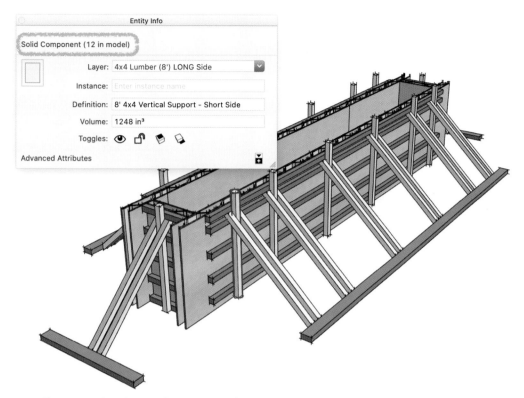

Figure 7.66   Quantity count for a component shown in the **Entity Info** inspection window.

- The second advantage is the ability to replace components with other components. This is especially helpful when the formwork is in the design phase or modifications are made during construction. For the component replacement procedure, see the step-by-step instructions shown in **Chapter 1—Components and Groups**.

- In the case where you want to replace one or two components from a series of components, you can make these components unique. Select the component or components you would like to make unique. With a right mouse click anywhere over the component or components, select the **Make Unique** option from the context menu. SketchUp will automatically assign a new component name in the following format **ComponentName#Numeral** (the numeral shows the number of unique components). Unique components can be reviewed in the **Components** inspection window (**Fig. 7.67**).

Figure 7.67 Applying the **Make Unique** options to a component. The **Components** inspection window will assign a new name to the unique component.

- The last advantage of using components is the ability to sustain a file that is manageable in size. Three-dimensional formwork models demand the use of a large quantity of repeating segments during the modeling work. Components are a perfect vessel for this type of work. For more information on the advantages that components offer, see **Chapter 1—Components and Groups**.

The organizational workflow, elaborated earlier, can be equally used for lumber or steel-type formwork. The only difference with the steel type of formwork is that you will have to get familiar with all of the different components of the formwork design, which can be done by reviewing the formwork shop drawings. The steel components in this formwork example consist of panels only. Assigning individual layers and creating individual components still apply, since revisions are a normal occurrence for any type of formwork during the construction process. Another benefit to remember is the fact that every component you make should be stored in your components library and can be accessed when modeling different portions of the project. One example would be clash detection between reinforcement bars and formwork accessories. See **Chapter 4—Introduction to Information Modeling and Organization** for further details on organization and how to generate detailed reports regarding the material type, quantity, price, and so forth.

When addressing complex formwork designs, the number one rule to remember is good organization, and everything else will fall in line!

CHAPTER 8

Modeling of Bridge Decks and Girders

Modeling of Noncomplex Bridge Decks	Modeling Jig
Modeling of Complex Bridge Decks	Fabrication Schedule
Elevation Points	Construction Schedule
Modeling of Girders	As-built and Revision Tracking
Steel Girders	

This chapter reviews the necessary steps required to model a three-dimensional representation of complex and noncomplex bridge decks and the underlying girders. Each bridge deck modeling workflow reviews what makes it complex versus noncomplex and how you can separate them in the initial review of the contract plans.

In addition, the last section of this chapter reviews the necessary steps to model concrete and steel girders. There will be a workflow example for a steel girder that was specifically chosen to show how you can use these three-dimensional models not only as visual aids but also as tools for initial and production planning purposes.

I do want to caution the reader that Chapter 8 sligtly differs from other chapters in this book, in terms of the detail of the step-by-step examples. Since there are a lot of different posibble shapes of bridge decks and different sizes of concrete and/or steel girders, it will not be beneficial to concetrate on one typical geometry. The idea behind Chapter 8 is to showcase the workflow and the required steps to be sucessful whenever you are tasked to model bridge decks and girders, regardless of the complexity or shape.

Modeling of Noncomplex Bridge Decks

Before the workflow review can start, you have to identify, for modelling purposes only, what makes a bridge deck a noncomplex type. From personal experience, in the noncomplex group you can classify any deck that has a constant cross-grade along the entire length and nonvarying width. Basically, any deck where you can draw a cross-section and extrude that cross-section to the correct length, without any major geometric changes, will fall into this category.

In order to model a noncomplex bridge deck, you need three types of very general information, which is readily available in the contract plans:

1. Typical cross-sectional view for the span of bridge deck you are planning to model.
2. Plan view of the span of bridge deck you are planning to model.
3. Finished deck elevations at each end of the bridge deck.

For this workflow example, use the general plan view and the cross-section of the bridge deck shown in **Fig. 8.1.**

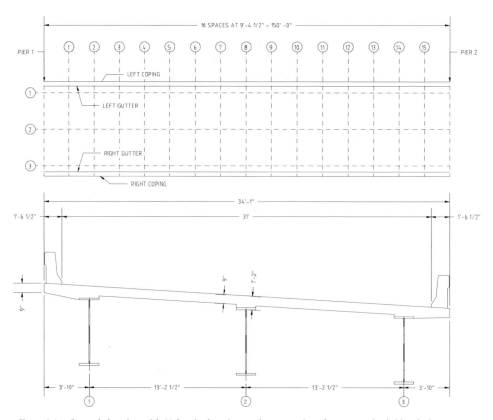

Figure 8.1   General plan view with 10th point locations and cross-section of a noncomplex bridge deck.

1. The first step in modeling the bridge deck is to draw the cross-sectional view. Model only the bridge deck and the haunch area. The parapets will be modeled separately in order to generate an accurate volume quantity of the deck and the parapets. Similar to the parapets, the bridge girders will also be modeled separately for accurate volume control, and they will be saved under a different layer in order to be used for later processing.

There are two options to draw the cross-sectional view of the bridge deck. One option is to import the AutoCAD file into SketchUp. After the importing is complete (following the information provided in **Chapter 3—Importing Files, AutoCAD, and SketchUp**), use the **Rectangle** tool, and create a surface that will overlay the entire cross-section of the bridge deck (**Fig. 8.2**). A mix of different assist lines will provide reference points when overlaying the drawing with the Line tool.

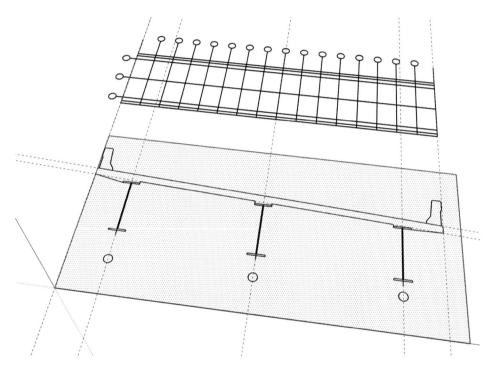

Figure 8.2   Overlaying a rectangular surface over the imported cross-sectional geometry of the bridge deck.

Modeling Tip

Toggle the visibility **ON** and **OFF** when tracing the outline of the deck cross-section by selecting the scenes buttons which are located below the toolbar. The ability to hide the base drawing can help you greatly during your modeling process. Review Chapter 3-Importing Files, AutoCAD, and SketchUp for more information on the proprt setup and usage of an AutoCAD file in SketchUp.

The second option is to model the bridge deck cross-section from "scratch," following the dimensions in the contract plans. Either option will work fine: just remember one option might be more time-consuming than the other.

2. Delete any excess surface or lines around the deck cross-section that were left over from the modeling process. After the excess lines and surfaces are deleted, check if the surface inside of the cross-section is still present by clicking on it with the left mouse button (**Fig. 8.3**).

Figure 8.3   Completed surface in the shape of the bridge cross-section.

3. The next step is to extrude the created cross-section of the deck. Since the example deck has no curvature, you can utilize the **Push/Pull** tool to extrude the deck (**Fig. 8.4**).

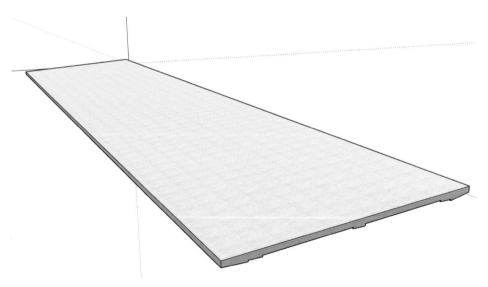

Figure 8.4   Three-dimensional representation of the extruded bridge deck based on the cross-section created earlier in the steps.

In the case of the bridge deck being curved, you will have to utilize the **Follow Me** tool. To do so, you will have to create an edge that will also provide a direction for the extrusion. In order to explain this situation bettwe, the following example and parameters will be used: the deck cross-section shown earlier in the workflow will follow a radius of 1500′, with an arch length of 150′. There are two ways that you can reproduce the radius of the deck. **Option 1** involves redrawing the outside radius of the span and **Option 2** involves tracing the radius of the span directly from the imported AutoCAD file. The procedure for each of these options is as follows:

Option 1:

a. Draw a circle with 1500-ft radius utilizing the **Circle** tool. Do not forget to increase the number of segments from the default number of 24 to 300. This is the only time I would recommend increasing the segment number (**Fig. 8.5**).

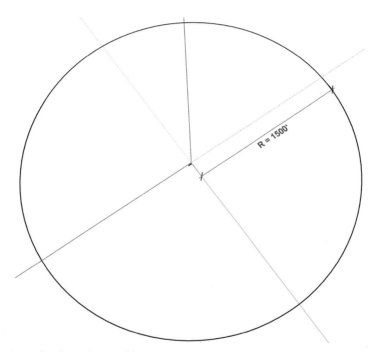

Figure 8.5   A 1500-ft radius circle created from 300 segments.

b. Calculate the internal angle based on the arc length of the span, from the example data, with a radius of 1500 ft and an arch length of 150 ft the internal angle is 5.7296 degrees. Trigonometric math is involved in this step; dust off the old college books!

c. Add an assist line to find the center of the circle. Rotate a secondary assist line based on the calculated angle (**Fig. 8.6**). The section between the two assist lines represents the radius of the deck span, which is directly associated with the arc length. Trim the excess edge located on the outside of the assist lines.

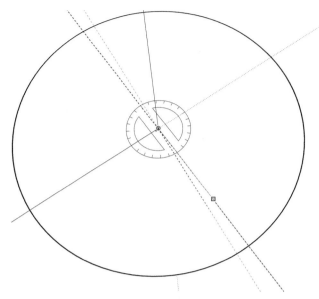

Figure 8.6   Creating the start and end points on the bridge utilizing assist lines, based on the curvature radius and arch length.

d. The last step is to move the created line to the end of the cross-section of the deck and use the **Follow Me** tool to extrude the curved bridge deck. One very important rule to keep in mind is that the curved line has to be tangent to the deck cross-section or you will end up with an inaccurate cross-sectional angle (**Fig. 8.7**).

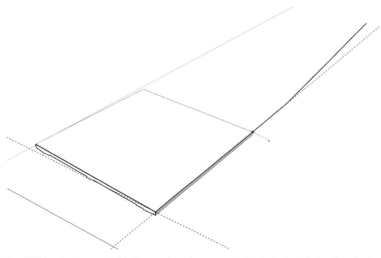

Figure 8.7   Utilizing the line created in the previous step as a path for the **Follow Me** tool in order to create a curved bridge deck.

Option 2:

a. Review the imported AutoCAD drawing of the plan view of the deck span section.

b. Utilizing the **Tape Measure** tool, set two assist lines at each end of the span.

c. Select the **2 Point Arc** tool and draw an arc, following the start point, end point, and the radius of the tracing drawing.

d. Select the newly drawn radius edge, and with the help of the **Entity Info** inspection window, verify that the **Radius** and **Arc Length** shown correspond to the information given in the imported drawing.

Modeling Tip

Most modelers will choose **Option 2** since it does not involve any trigonometry in order to create the radius of the span. A word of caution: always check the accuracy of the drawing that you intend to import and use as the base for tracing. Drawings that are not printed in the correct scale, or content that is not in the correct scale during drafting procedures, can create substantial problems in your three-dimensional model.

---

e. Select the entire three-dimensional representation of the deck span. Click on the model with the right mouse button and from the context menu select **Make Group (Fig. 8.8)**.

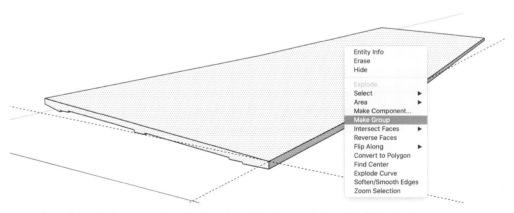

Figure 8.8   Creating a group from the three-dimensional representation of the bridge deck.

Modeling Tip

Unlike other chapters where you make components of the three-dimensional models, for the deck cross-sectional drawing a group is more adequate since you do not anticipate making multiple copies. Each span of a deck section is almost its own entity. Review **Chapter 1—Components and Groups** for more information on the properties of groups and components.

Modeling of Complex Bridge Decks

Bridge decks with varying cross-slopes, steep superelevation transitions, and varying widths, from a visual perspective, can be classified as complex bridge decks. Most of the previously mentioned properties prevent the user from creating a default cross-slope and extruding it across the longitudinal length of the bridge, as you did in the **Modeling of Noncomplex Bridge Decks** example. Complex bridge decks involve significantly more planning and three-dimensional modeling to create the correct finished deck geometry.

In order to model a complex bridge deck, you need the following information, which is readily available in the contract plans:

- **Finish Grade Elevation**—Provides the elevation point at every "tenth" point along the deck length. The information is provided for each girder line, gutter line, coping, baseline, and the like.

- **Superstructure Plan**—The superstructure plan will provide the user with an overall plan view of the deck. The information provided in the superstructure plans includes arc geometry, radius of the edge of deck, location point of varying width, overall longitudinal length of the deck, and so on.

- **Superstructure Sections**—Similar to the superstructure plans, the superstructure section provides information on the overall width of the deck at certain key locations along the bridge. Besides the overall width of the deck, it also provides spacing of lanes, breakdown lanes, construction stages, and so forth.

- **Framing Plan**—Although not always necessary, it can be a source of information for the curvature radius of girders, overall spacing of girders, longitudinal length of spans, and the like.

For this workflow example use the finish grade elevation table, the plan view, and the cross-section (**Fig. 8.9**). This is the same cross-section as you used in the previous example.

Finish Grade Elevations - Span 1

Location	Pier 1	1	2	3	4	5	6	7	8	9	10	11	12	13	14	15	Pie
Left Coping	36.45	36.06	35.65	35.26	34.91	34.56	34.22	33.89	33.57	33.26	32.96	32.68	32.40	32.14	31.89	31.65	31.
Left Gutter	36.42	36.02	35.61	35.23	34.88	34.53	34.19	33.86	33.54	33.23	32.93	32.65	32.37	32.11	31.86	31.62	31.
Girder 1 (Center Line)	36.35	35.96	35.56	35.18	34.83	34.48	34.14	33.81	33.49	33.18	32.89	32.60	32.33	32.06	31.81	31.57	31.
Girder 2 (Center Line)	35.98	35.63	35.28	34.92	34.57	34.22	33.88	33.55	33.23	32.92	32.62	32.34	32.06	31.80	31.55	31.31	31.
Girder 3 (Center Line)	35.61	35.31	34.99	34.65	34.30	33.95	33.61	33.28	32.69	32.66	32.36	32.07	31.80	31.54	31.28	31.04	30.
Right Gutter	35.55	35.25	34.94	34.61	34.26	33.91	33.57	33.24	32.92	32.61	32.31	32.03	31.75	31.49	31.24	31.00	30.
Right Coping	35.51	35.21	34.91	34.58	34.23	33.88	33.54	33.21	32.89	32.58	32.28	32.00	31.72	31.46	31.21	30.96	30.

Figure 8.9   Finish grade elevation table at 10th point locations: general plan view with 10th point locations and cross-section of a complex bridge deck.

1. Unlike the previous example, the first step in the workflow is to start creating the plan view of the bridge span. We cannot draw the cross-section since the deck grades change and thus you will create an incorrect surface that will not match the finish grade elevations. There are two ways to accomplish this task, you can either

import an AutoCAD drawing and trace the plan view utilizing surfaces, assist lines, and so on or create the plan view from scratch. Either option works fine as long as you set up and import your AutoCAD file correctly. For this example, you can draw the plan view from scratch since it is very simple (**Fig. 8.10**).

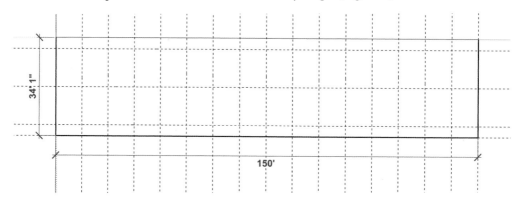

Figure 8.10 Plan view of the bridge deck with 10th points, gutter lines, and coping marked with assist lines.

2. Select the **Line** tool and starting from left to right, create elevation point, by drawing lines on the **Blue** axis. Each vertical line is created based on the elevations at each 10th point. In order to make your life easier you can create a datum line on the same elevation as your plan view and draw the elevation differences between the 10th points, checkout the Modeling Tip below for more information. The final result should look similar to **Fig. 8.11**.

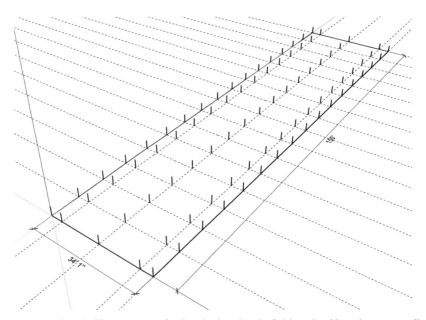

Figure 8.11 Each vertical line represents the elevation based on the finish grade table at the corresponding 10th point along the bridge deck.

Modeling Tip

The elevation points shown in the **Finish Grade** table are based on real-world data. Choosing a baseline that will encompass the lowest elevation point in the series will save the user a lot of unnecessary time exhausted on scrolling between the elevation zero mark and the top elevation point of the deck.

3. The next step is to connect the elevation points in the longitudinal direction of the bridge deck. The tool you choose for this task depends on whether the bridge has a curvature or not. For this example, the bridge deck has no curvature; therefore, use the **Line** tool to connect the elevation markers created in step 2. In cases where the deck has a curvature use the **2 Point Arc** tool, and connect three points at a time. The final result should be similar to **Fig. 8.12**.

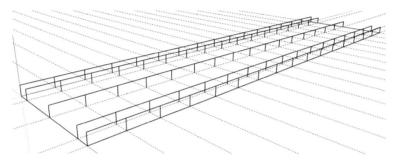

Figure 8.12   Elevation markers connected utilizing the **Line** tool.

Modeling Tip

Step 3 of the exercise combines the creation of the bridge grade and the overall longitudinal curvature into one task.

4. Utilizing the **Move** tool, create offset copies of the two outside bridge deck lines. Select both outside bridge deck lines, click on the **Move** tool, and by holding down the **Option** key, create an offset copy. Type the total offset distance of 9.0 in. in the **Measurements Input Box** in order to create copies with accurate offset distance (**Fig. 8.13**). The offset lines represent the thickness of the bridge deck.

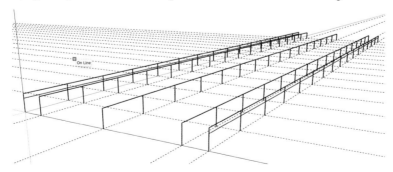

Figure 8.13   Creating the thickness of the deck by making an offset copy of the bridge coping lines.

5. The next step is to create surfaces between all the lines created in step 3. For this purpose use the **From Contour** tool that is part of the **Sandbox** toolset. Select all lines, except the offset lines, assist lines and click on the **From Contour** tool. SketchUp will generate surfaces between all the created lines (**Fig. 8.14**). (The Sandbox toolset is discussed further in **Chapter 9—Site Modeling and Use of Site Creation Tools.**)

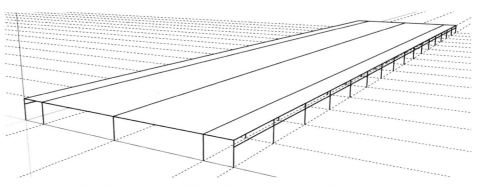

Figure 8.14 Creation of the deck surface utilizing the **From Contour** tool in the **Sandbox** tool set.

Step 5 completes the creation of a complex deck. The next set of logical steps would involve the creation of the underside of the deck and each side utilizing the two lines that were copied and offset in step 4. These steps are not done as a workflow example since the example was designed more to show the modeling work of the deck itself, and you can complete the rest of the model using the tools already discussed. Do not forget to make a group of the deck when you complete the model.

The preceding workflow example creates a complex bridge deck, with no assumptions of stage construction. Because of this assumption, you created a single group that will generate the volume and therefore the concrete quantity as a unit.

In cases where the bridge deck is constructed in several stages, each stage will need to be modeled separately. Furthermore, separate layers and separate groups will need to be created for each stage. By doing so, at the end of the modeling work you will have access to the different attribute information based on each step and stage of construction. This is specifically important when a report is generated for further use; the bridge deck will be correctly divided. Furthermore, creating separate layers and groups of the deck also gives you the ability to manipulate with the visibility options, and therefore tailor them per the overall construction schedule.

Modeling of Girders

The general idea behind this chapter is not to concentrate heavily on the modeling aspect of girders, but to provide an understanding of how these girder models fit in the information modeling concept and can ultimately help you in all aspects of design and construction work. Three-dimensional modeling should be equally helpful in the information aspect as well as be visually appealing.

Commonly on bridge projects, girders are either prestressed or post-tensioned style concrete girder or steel plate girders. The prestressed or post-tensioned girders are very easy to model since they either fall into the singly symmetrical category or doubly symmetrical category. A workflow example is not included since it is similar to the pile modeling workflow that was completed in **Chapter 5—Modeling of Substructure Components**. This section concentrates on reviewing the necessary workflow to create an accurate three-dimensional representation of steel plate girders since they are more informationally complicated.

Steel Girders

The most commonly used steel girder types on bridge projects today are the I-shaped girder and the U-shaped girder, also known as tub girders. For the purpose of this modeling example, a U-shaped steel girder will be reproduced, with dimensional properties as shown in **Fig. 8.15**.

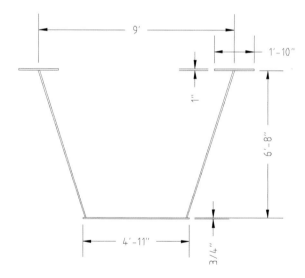

Figure 8.15   Cross-sectional shape of a U-shaped steel girder for workflow example purposes.

The idea behind the workflow is to show the correct steps to complete the modeling portion of the work efficiently. Furthermore, there will also be a review of the types of additional information you can gather from the model, besides being a visual assistant. All steel girders are created from different size plates, welded together. Flange plates can vary from span to span, splitting the span, or at specific section that may be located over piers, among others. Before starting the workflow example, you need to review several key points to modeling steel plate girders:

- Review the contract plans or shop drawings to understand the locations where the girders change plate sizes. Shop drawings are more accurate to review since fabricators will revise certain portions for ease of manufacturing. If the model is made for estimating purposes, and therefore no shop drawings are available, review the girder elevation drawings in the contract plans.

- Review the overall geometry of the girders utilizing the framing plans, specifically the curvature radius, if present.

- Always draw the cross-section of the girder as one continuous face. This point might be counterintuitive since as the name says, steel plate girders are made from individual plates and therefore you would think to model them as such. This will be discussed later in the section as to why this is the case and what you gain when you model them as one continuous face.

- Locate the different types of internal braces, cross-braces, stiffeners, and so on.

- Locate the specific areas where steel with different specifications is used. Color-coding these areas can be of great benefit during estimating purposes.

- Make notes of protrusions in the girders. These protrusions can be in the form of holes for inspection access, drainage pipes, fire suppression pipes, ITS conduits, and so on. Subtracting the steel from these areas can make a difference when erection work plans are made.

You can put these key points into practice and start the modeling workflow of the U-shaped steel girder shown in **Fig. 8.15**.

1. The first step is to create the cross-sectional face of the U-shaped girder. Select the **Rectangle** tool and create a rectangle face with the following dimensions 81.75 in. × 130 in. (**Fig. 8.16**).

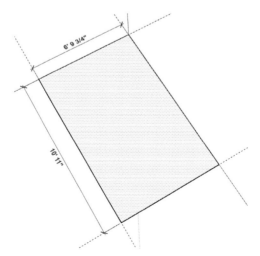

Figure 8.16   Rectangle face created to engulf the entire cross-section of the U-shaped girder.

2. Select the **Tape Measure** tool and place two horizontal assist lines to mark the flanges as shown in the above **Fig. 8.15**.

3. Select the **Tape Measure** tool again, and this time place vertical assist lines to mark the center and end points of the top two flanges and the bottom flange (**Fig. 8.17**). Follow the dimensions shown in the above **Fig. 8.15**.

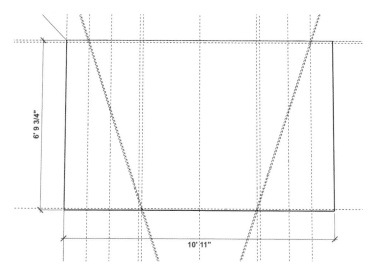

Figure 8.17   Assist lines used to mark the outline of the U-shaped steel girder on the rectangle face.

4. Utilizing the **Line** tool, draw in the outline of the cross-sectional shape of the steel girder (**Fig. 8.18**).

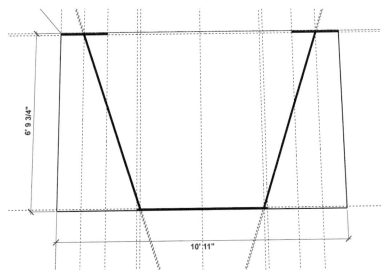

Figure 8.18   The outline of the U-shaped steel girder is completed utilizing the Line tool and the assist lines created in the previous step.

At this point, in the workflow, the next step would be to add the longitudinal component of the girder. In order to do that, you will need to determine if the girder has a curved or straight component by reviewing the framing plan drawings. For this particular example, the girder will not have a radius in order to shorten the length of the example but still be able to show the key points in the modeling a steel girder steps. The idea behind this example is to review the benefits of three-dimensional models of a beam, or in this case a girder. The end of this section concludes with instructions on how to create longitudinally curved girders for anyone who is brave enough to try!

5. Select the cross-sectional shape of the girder and make a copy. Select the **Move** tool and press and hold the **Option** button on your keyboard. Click on any corner of the cross-sectional shape, and create a copy (**Fig. 8.19**). You will use this copy for future modeling work of the splice plates and other internal stiffeners.

Figure 8.19 Initial cross-section and offset copy of the U-shaped steel girder. The offset copy will perform a "jig" purpose for the creation of other steel components.

6. Select the original cross-sectional shape and utilizing the **Push/Pull** tool, extrude 60 ft (**Fig. 8.20**).

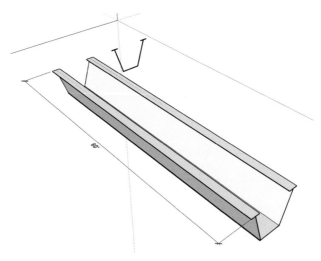

Figure 8.20 Extrusion of the steel girder.

In the next steps you will create a separate layer in order to organize the models properly for further use.

7. From the **Layers** inspection window (**Tags** in SketchUp version 2020), add two layers by selecting the plus (+) icon located in the upper-right corner (**Fig. 8.21**). Name the layers **Segment 1** and **Working Section**.

Figure 8.21   Addition of the layers **Segment 1** and **Working Section** to the **Layers** inspection window (**Tags** in SketchUp version 2020). Layer will be used for future work and to add an informational component to the three-dimensional model.

8. Select the segment and move the content to the **Segment 1** layer, by selecting the **Layer** drop-down menu located at the **Entity Info** inspection window (**Fig. 8.22**). Repeat the same steps for the other two segments.

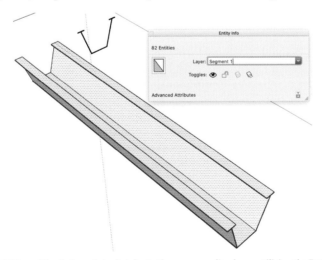

Figure 8.22   Addition of the U-shaped steel girder to the corresponding layer, utilizing the **Entity Info** inspection window.

9. Select the cross-sectional segment and move the content to the **WorkingSection** layer, by selecting the **Layers** drop-down menu located in the **Entity Info** inspection window (**Fig. 8.23**).

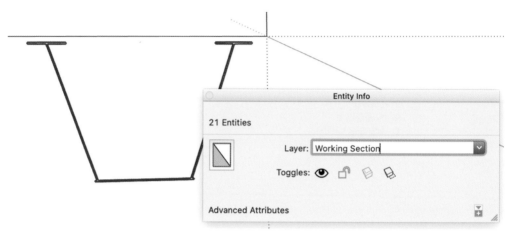

Figure 8.23 Addition of the copied U-shaped steel cross-section to the corresponding layer, utilizing the **Entity Info** inspection window.

10. The last step is to create components for the girder segment and the cross-sectional face that you copied in step 5. Select the girder segment and from the context menu select the **Make Component…** option. The context menu can be activated with a right mouse click anywhere on the segment.

11. In the **Components** inspection window, record the segment name and location in the **Definition** and **Description** text boxes.

12. The main option that you have to pay attention to is the **Set Components Axis** option. At every splice location for steel girders, designers leave a small gap from girder segment to girder segment. This gap can vary based on design requirements, manufacturing tolerances, state regulations, and so on. In order to account for this gap, offset the local axis in the same dimension as the gap. Click on the **Set Components Axis** button, which will change the mouse pointer from an arrow to the axis symbol. Hover over the centerline point of the bottom flange so that SketchUp's inference engine can provide you with a center point. Slowly hover away from the girder with the mouse, for the total distance of the gap. Left-click on the new point three times to set the new local axis for the steel segment (**Fig. 8.24**).

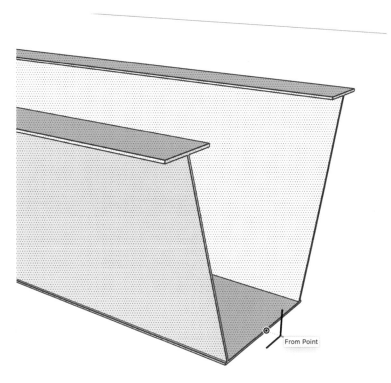

From Point

13. Repeat Step 11 for the cross-sectional segment. Name the segment **WorkingSegment** in the **Definition** text box. No need to write anything in the **Description** portion or to change the local axis. The cross-section will be used only as a modeling jig in order to help you during the modeling work of splice plates, stiffeners, diaphragms, and the like.

The modeling portion of the exercise is now complete. Congratulations!! The next step is to review what you have done, what information you can gather from the model, and what are the next steps.

Modeling Tip

The purpose for the cross-sectional face segment is to act as a "modeling jig" for future modeling work. One example would be when conducting modeling work for the internal cross-frames and stiffener plates (**Fig. 8.25**). Since the cross section face is a component, you do not have to worry about the "stickiness" property of edges and faces that is present in SketchUp.

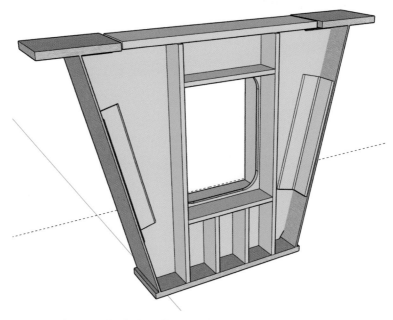

Figure 8.25   Modeling example of a three-dimensional representation of an intermediate stiffener created on the modeling jig.

- **Internal Components.** Every time an internal segment (cross-frame or stiffener plate) is modeled on the "modeling jig," convert it into a component and assign a specific layer. There are three main reasons for this. The first reason is that components give you the option to fully exchange one component with another. This property can be very helpful if revisions are made to the internal structure of the girder. The second reason is the fact that components take reasonably less file space when copied multiple times than do groups or just models. For more information on groups and components, review **Chapter 1—Components and Groups**. The last reason is that layers give you the opportunity to toggle the visibility **ON** and **OFF** for specific components, which can be a great benefit for keeping track on the fabrication or construction schedule—this will be talked about more in the last two bullet points.

- **Naming Convention of Internal Components.** The naming convention for the internal girder components is very important. In reality, each girder segment will have multiple internal components that graphically look the same. In order to not get confused and overwhelmed, always use the naming convention for these segments as shown in the shop drawings.

- **Color by Steel Specification.** As was mentioned earlier in this chapter, steel girders are often fabricated from several different steel types for structural reasons. The ability to distinguish between these different material segments is helpful for construction purposes, especially if you have to apply welding work. The workflow shown in **Chapter 8** creates **Layers** for each segment of the steel girders; in the example you only had one, but on a real project you will have multiple segments.

Each layer can be assigned a specific color, based on material type and used with **Color by Layer** option when needed for review.

- **Volume and Weight Data.** The reason for creating each girder segment from one continuous face is that it gives you the overall volume data of that segment. This is very important since you can convert that volume data into weight, by multiplying it with the correct material density (**Fig. 8.26**). This is particularly important for estimating purposes, when you need to calculate the total amount of steel material used in the girder structure or when you need to estimate the size of crane equipment for erection purposes, during construction. The volume only works if the model is made from one continuous face and is converted into a component.

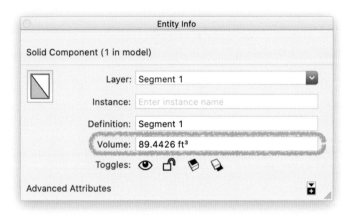

Figure 8.26   **Entity Info** and the **Volume** data of a particular segment.

- **What to Do with Different Thickness Plates.** Girder segments are fabricated from plate members, and most often these plate members change in thickness. This change in thickness does not always start or end within a particular girder segment. The workaround is to create different segment components in order to account for the change in plate thickness and at the same time be able to receive volume data.

- **Summing Everything Together.** The last point is to combine all the components together. In order to explain this particular workflow, easely, is best to examine only one segment. After creating the segment component, the next step is to review if the segment has any particular protrusions—this will make a difference in the overall volume and thus the weight of the member. If any protrusions are present, create them now by accessing double clicking over the segment and the component. After all the protrusions are completed, utilizing the **Tape Measure** tool, add assist line for the location of the internal cross-frame structures. As was mentioned earlier, all the internal cross-sections should be modeled on the modelling fig and converted into components. Add these components at the locations of the assist lines. Similar to the cross-sections, model on the modeling jig and convert to components the splice plates. Add splice plates to one end of the girder segment (**Fig. 8.27**). Figure 8.27 shows a typical, I-shaped girder with different components attached to it as specified in the steel shop drawings.

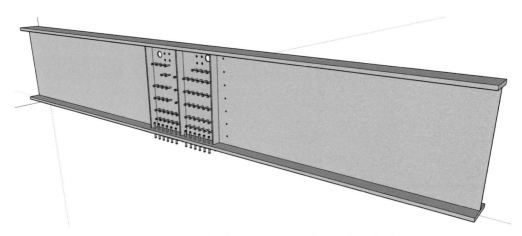

Figure 8.27   Representation of a steel girder with protrusions, shear studs, and stiffener plates.

- **Fabrication Schedule.** Besides other usages which were mentioned earlier, the three-dimensional model of the girder can also be used for keeping track of the fabrication schedule. This can be accomplished by toggling the visibility option **ON** and **OFF** from the **Layers** Inspection window (**Tags** in SketchUp version 2020) for each layer of a girder segment. This information can be exported and distributed to everyone on your team on a daily or weekly basis for review purposes.

- **Construction Schedule.** Similar to the **Fabrication Schedule**, the construction schedule can be monitored by toggling the visibility option **ON** and **OFF** from the **Layers** inspection window (**Tags** in SketchUp version 2020) for all the components that were erected in a particular work shift. This is a tremendous help for the manager and engineers as well as for field personnel who can review their own progress and effectiveness.

- **As-built and Revision Tracking.** During construction work, revisions are a common occurrence for many reasons, that is, for oversights during design work or construction inconsistencies. Whatever the case is, you will have to keep track of the different types of changes so you can plan out any future conflicts that can occur because of these changes. The first step in the workflow for tracking revisions and as-builts is to model the specific change, that is, internal or external cross-braces, stiffeners, protrusion locations, and so on by utilizing the "modeling jig." Create a component from the revised section, and utilizing the components options, fully replace one component with the other to make the required switch. For more information on groups and components, refer to **Chapter 1—Components and Groups**.

As you can see from above, there are ten different benefits that you can gain if you properly plan and model your girders and the superstructure portion of the bridge in general. These benefits can increase the productivity of your estimating team and at the same time extend all the information, without remodeling work, to the construction team in the field. Hope this model opened your horizons to the endless possibilities that SketchUp can provide.

CHAPTER 9

Site Modeling and Use of Site Creation Tools

Importing of Geo-location Data
    Add Location Window
    Constructing of Contour Lines

Introduction to Sandbox Tools and Terrain Creation
Three-dimensional Terrain Modeling
Utilization of Terrain Models

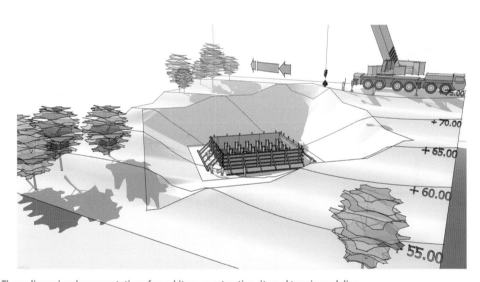

Three-dimensional representation of an arbitrary construction site and terrain modeling.

There are many options to use when creating accurate site models that can be later used for estimating or construction purposes. Some of these options require very little effort, as is the case when importing *.dem or *.kmz files or if you use SketchUp's own set of tools **Add Location…** and **Toggle Terrain**. Other options include more effort, as is the case when creating site models from scratch by utilizing SketchUp's **Sandbox** toolset or by importing CAD-type files and retracing the context. Another viable option is an additional third-party extension that offers the ability to import data points and create a site model. One thing to remember is that different import options also impact the accuracy of the final site model.

Before discussing the proper workflow for importing and using the site data, you need to understand the difference in data-type accuracy and for what purposes you want to use that data. The accuracy of site data depends on many factors, the following is a list of some of the more important factors to keep in mind for future use:

- The method used to capture the data—was the data captured by a satellite, airplane, drone, or conventional survey methods—different capture methods will offer different levels of accuracy.
- What is the density of the collected data and what is the pixel size or grid resolution—similar to the first point, this point is also in reference to the site model accuracy.
- The overall accuracy and the type of algorithms used for interpolation purposes during the analysis of the collected data—this point is more important for data gathered by aerospace means, satellites, planes, drones, and the like.

Making the initial decision for the level of data accuracy you want to use in your model depends on the purpose for which you want to use the data. Some examples for this are design, initial estimating purposes, advanced estimating purposes, construction work plans, construction checks, and so on.

The following three sections review the methods for creating a three-dimensional terrain model utilizing SketchUp's native tools. The workflows will showcase one example of using the standard **Add Location…** and **Toggle Terrain** tools available in SketchUp. The second workflow showcases the use of the **Sandbox** toolset to generate a terrain model, and the last section showcases for what purposes you can use the created three-dimensional terrain models.

The use of a third-party extension or the option to import CAD drawings was not reviewed. There are a large number of third-party extensions on the market, and different modelers have different requirements for what they see as useful in an extension. If you are interested in using a third-party extension, do some research and try several different options to see what works best for your modeling need. Furthermore, the option to import CAD drawings for terrain modeling only requires tracing skills to accomplish, so this option is not reviewed.

Importing of Geo-location Data

One useful options to create terrain, which also comes standard with the SketchUp application, are the **Add Location…** and **Toggle Terrain** tools. A specific location is used for example purposes—Horseshoe Falls, which is located in Niagara Falls State Park. This location was chosen on purpose in order to visually show the pronounced and abrupt changes in elevation, which in turn will help better explain the underlined workflow.

To get started and create a three-dimensional terrain model, check to make sure the following tools are enabled: **Add Location**, **Toggle Terrain**, and **From Contours** (**Fig. 9.1**).

Figure 9.1   Icon representation of the tools needed for terrain modeling.

If they are not enabled, please do so, since you will use them heavily for this example. If you do not remember how to add tools, refresh the steps outlined in **Chapter 1— Introduction to SketchUp**.

1. From the toolbar, left-click on **Add Location…**, which will open the **Add Location** window. Before you continue to step 2, this is a good stopping point to review the **Add Location** window.

Add Location Window

The main parts of the **Add Location** window are as follows (**Fig. 9.2**):

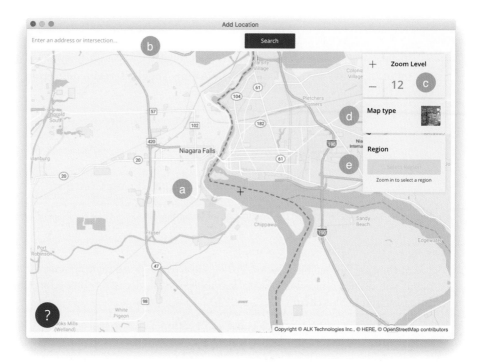

Figure 9.2   View of the **Add Location** window.

a. **Search Region**: Allows you to search for a specific map location utilizing the mouse scroll functions.

b. **Search Box**: Allows you to search for a specific map location that will be used in a three-dimensional model. Location can be searched by physical address or a coordinate system. When searching in coordinate system, the location can be entered in a decimal format (43.077548, −79.074567) or in degree, minutes, seconds format (43°04′39.17″N, −79°04′28.44″E).

c. **Zoom Level**: Allows you to change the zoom level; the larger the zoom level number that is chosen in the option box, the more accurate the imported data will be in the model.

d. **Map Type**: Allows you to toggle between satellite and map view.

e. **Region**: Allows you to select the region you want to import for modeling purposes.

2. The next step is to search for the location you want to import. In the search box type the following address: Horseshoe Fall, ON, and press the **Search** button (**Fig. 9.3**). As mentioned before, you can also enter the address in the following nomenclature: 43.077548, −79.074567 or 43°04″39.17″N −79°04″28.44″E.

Horseshoe Falls

43.077548, -79.074567

43° 04′ 39.17″ N -79° 04′ 28.44″ E

	Search
	Search
	Search

Figure 9.3   Different options to enter a location in the search field.

Modeling Tip

When entering a location in the coordinate system format, care has to be taken so that space is left between the two coordinate numbers.

3. Set the zoom level to 17 either by zooming in or out utilizing your mouse wheel or by using the plus or minus buttons next to the zoom (**Fig. 9.4**).

Figure 9.4   The zoom is set at the corret zoom level of 17 for the geo-location.

4. When you are satisfied with the location Click on the **Select Region** button. The **Select Region** button will turn into the **Import** button. Click on the **Import** button and your selection will be imported into the model (**Fig. 9.5**).

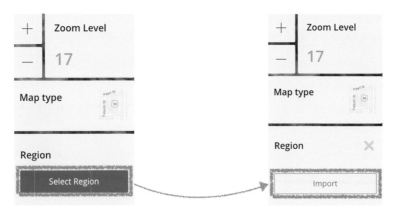

Figure 9.5   Final steps of the procedure to import a location into SketchUp.

Modeling Tip

The border around the imported location can be adjusted further by clicking and dragging on one of the four white dots (**A**) that are located on each corner of the import box (**Fig. 9.6**). This has to be done prior to pressing the **Import** button.

Figure 9.6   Ability to adjust the import window by movement of the circular points located on each side of the import window.

Constructing of Contour Lines

The next step, after importing the location block of the terrain, is to create a contour map by placing contour lines. The elevation at which the contour lines will be placed depends on two key elements: how drastic the elevation change of the terrain is and for what purpose you will use the map. On steep terrains with a consistent rise in elevation, contour lines can be placed every one hundred to two hundred feet. This rule of thumb can also be used for preliminary estimating work. In cases where the terrain model will be used for logistics purposes (moving of material and equipment), the contour lines should be placed at every two to five feet, in order to get a visual feel of the terrain and estimate if it falls within the recommended guidelines for traversable slopes.

1. From the toolbar, click on **Toggle Terrain** tool, which will reveal the elevation changes in the imported section of the map (**Fig. 9.7**).

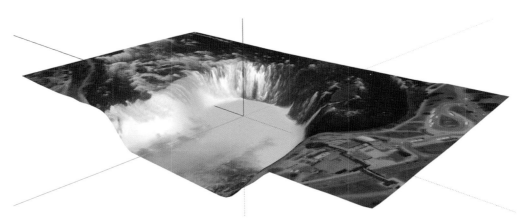

Figure 9.7   The imported map with the terrain activated by clicking the **Toggle Terrain** tool.

2. By default, SketchUp places the imported map at the center of the global axis. In order to estimate the total elevation of the imported map, select the **Tape Measure** tool, and offset an assist line from either the **Green** or **Red** axis until it touches the top elevation of the map (**Fig. 9.8**).

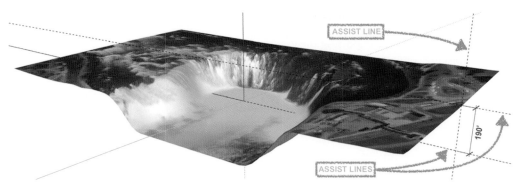

Figure 9.8   Addition of assist lines with the help of the **Tape Measure** tool, in order to help with the very rough estimate of the terrain elevation.

3. From the toolbar, select the **Rectangle** tool and draw a rectangular surface that is slightly larger than the imported map. The rectangle should be placed at the lowest elevation level of the imported map (**Fig. 9.9**).

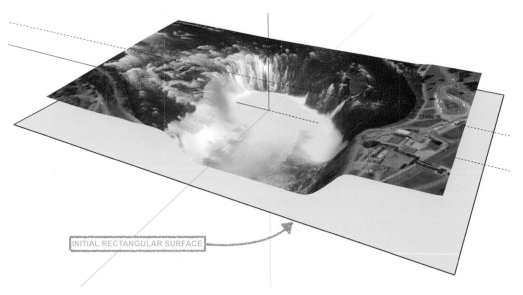

Figure 9.9   Creation of the initial rectangular surface that will be later used to create the contour lines.

4. Based on the example map, the total elevation difference is about 190 ft (**Fig. 9.10**). You do not have to be very accurate with the elevation, since you will only use it to estimate the elevation difference between the contour lines.

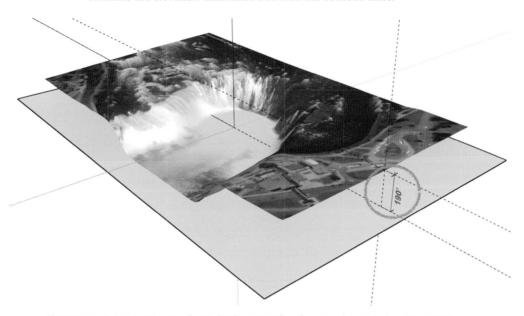

Figure 9.10   Estimating the overall scale for the contour lines based on the rough elevation of 190 ft.

5. Select the rectangular surface, and make an offset copy of 10 ft by utilizing the **Move** tool and holding down the **Option** button of the keyboard. Do not forget to select the upper arrow key from the keyboard in order to lock the movement of the offset copy to the **Blue** axis (**Fig. 9.11**). For example purposes, use 10 ft for an elevation difference between the contour lines.

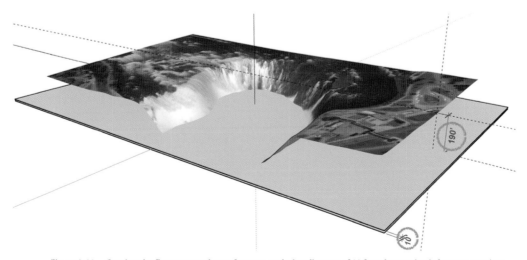

Figure 9.11   Copying the first rectangular surface at a evelation distance of 10 ft and preparing it for array copying.

6. Without clicking on anything else, type **21´** in the **Measurements Input Box** and press Enter. SketchUp will make an additional 20 copies, counting the original offset copy, spaced at 10 ft on center (**Fig. 9.12**). By copying the rectangular surface you just created all your contour lines.

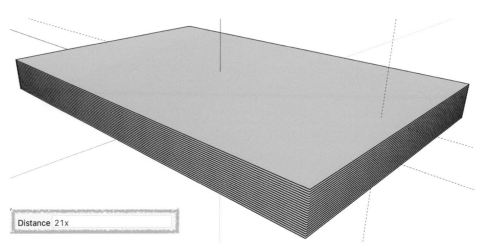

Distance 21x

Figure 9.12   Array copying of the rectangular surfaces.

7. Select all the rectangular surfaces and the imported map, and activate the context menu by clicking with the right mouse button. From the context menu, hover over **Intersect Faces** and select **With Model** (**Fig. 9.13**).

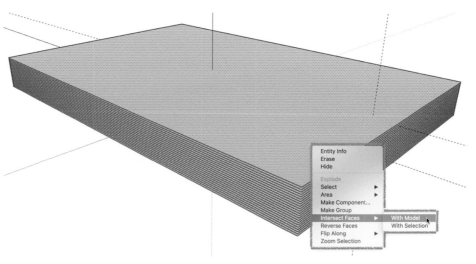

Figure 9.13   Creating intersection points between the rectangular surfaces and the imported terrain using the **Intersect Faces** option from the context menu.

8. Select all the rectangular surfaces and delete them. You are left with a contour map that is an imprint of the imported terrain (**Fig. 9.14**).

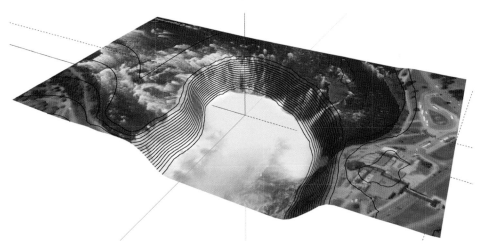

Figure 9.14   Contour lines are imprinted into the imported map.

At this portion of the workflow exercise, you can use the contour map as shown in **Fig. 9.14**, with the terrain attached to it, or you can go one step further and create a contour map with surfaces as shown in steps 9 and 10.

9. The first step is to create a layer that will be associated with the contour lines. Creating a layer will also come in handy when you create the terrain surface. From **Layers** dialog box (**Tags** in SketchUp version 2020), select the plus sign (**A**) to create a layer. Name the layer **Contour Lines** (**B**) and change the line type to dashed (**C**) by clicking on the line symbol and selecting the appropriate line (**Fig. 9.15**).

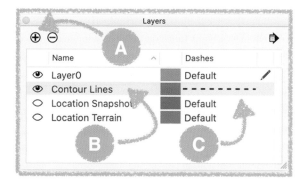

Figure 9.15   The **Layers** inspection window allows you to change the line type.

10. The next step is to delete the imported terrain (it is no longer required for the model) and move the geometry from the default layer, **Layer0,** to the newly created layer, **Contour Lines**. Complete this step by select all the geometry and then select the **Contour Lines** layer located under the **Layer** drop-down menu in the **Entity Info** dialog box (**Fig. 9.16**).

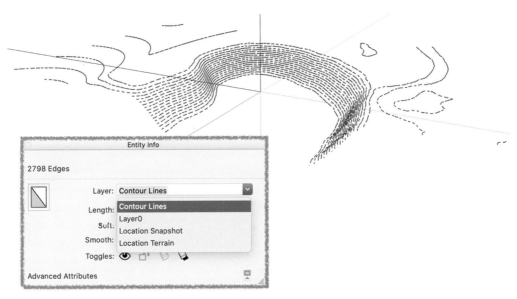

Figure 9.16   Moving the contour line geometry from the default layer to the newly created **Contour Lines** layer, using the **Entity Info** inspection window.

11. The next step is to give the contour lines a frame. By framing the geometry, you are helping the **From Contours** tool connect all the edges and giving a more accurate representation of the terrain. **Figure 9.17** shows a surfaced terrain that did not have the proper border completed versus a surfaced terrain where the border was properly completed.

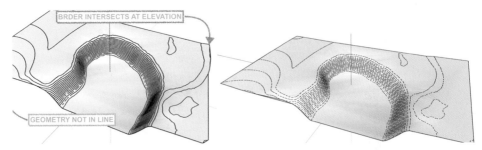

Figure 9.17   Side-by-side comparison of a three-dimensional representation of a terrain with completed and not completed border.

In order to create a border, select the **Line** tool and fill in between the points of the counter lines. In areas where the counter line points are scarce, use a combination of assist lines by utilizing the **Tape Measure** tool in order to give yourself a direction of work (**Fig. 9.18**).

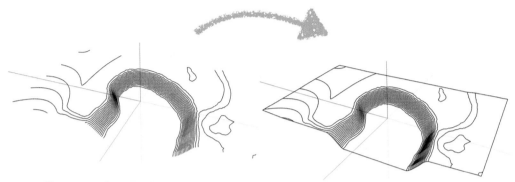

Figure 9.18   The outlined border is completed for the terrain model.

12. Select all the contours and click on the **From Contours** tool, which will create surfaces (**Fig. 9.19**).

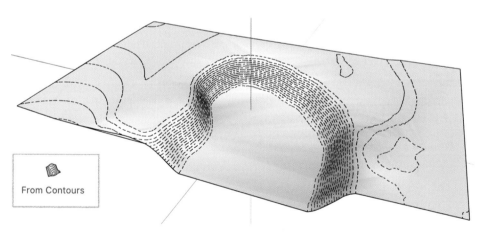

From Contours

Figure 9.19   Creating a surface between the contour lines by using the **From Contours** tool.

This concludes the workflow exercise on how to create three-dimensional terrain by utilizing standard SketchUp tools. In the next two sections, you will create a three-dimensional terrain utilizing the native SketchUp toolbar called **Sandbox** and explore how its toolset can be used in the heavy civil industry.

Introduction to Sandbox Tools and Terrain Creation

The following section reviews the SketchUp native toolset called **Sandbox**. Furthermore, it reviews the workflow of how to create and modify a three-dimensional terrain model with the assistance of the **Sandbox** toolset through modeling examples. The **Sandbox** toolset consists of seven individual tools, all with different modeling properties: **From Contour**, **From Scratch**, **Smoove**, **Stamp**, **Drape**, **Add Detail**, and **Flip Edge** (**Fig. 9.20**).

Figure 9.20   Icon representation of the **Sandbox** toolset.

Something that is important to note is that the **Sandbox** toolset should not be viewed as a terrain-specific tool. The **Sandbox** toolset can be used for many other complex surface applications, as was the case in **Chapter 8** when reviewing the workflow example for modeling complex bridge decks.

Three-Dimensional Terrain Modeling

The following workflow example reviews the necessary steps to develop a three-dimensional terrain model. For this purpose use the following tools: **From Scratch, Smoove,** and **Add Detail**. This option of creating terrain models is significantly more complex than the option that was reviewed earlier, where we imported a section of terrain.

There will be no files associated with this example. The example will be created as a freehand exercise; therefore, you can organize the terrain as you wish while following the steps in the exercise.

1. The first step in the workflow is to create a base grid, created from a series of surfaces. Before starting the modeling work, you need to consider a couple of factors when creating a grid:

   a. Review the project terrain drawings and estimate what size of grid you need. For this example use a grid size of 100 ft × 100 ft.

   b. Take a note on the overall geometry of the terrain, specifically any topographic features that are part of the terrain, that is, hills, valleys, or other smaller depressions.

   c. Size of the components that compose the grid area. The initial tendency is to create a grid containing very small components. This tendency arises from logic that dictates that smaller grid components equal higher accuracy in the terrain model. Although this is a true statement, this logical thinking can also grind your modeling experience to a stop. SketchUp generates the grid area as a series of surfaces interlinked with each other. This is not a serious issue when working with small grid areas. The issue comes to light when working with large grid sizes, for example, 10,000 ft, × 10,000 ft, or 10,000 ft, × 10,000 ft, and when you choose a grid component of 1 ft or 6 in. If you do a quick calculation, SketchUp will have to generate 100,000,000 individual surfaces for a 10,000 ft, × 10,000 ft, grid, spaced at 1-ft increments. In order not to lose any accuracy but at the same time keep the modeling work to a manageable level, SketchUp allows the user to subdivide specific areas to smaller increments after the grid is created; this is discussed further in the example. For this example make a grid of 5-ft × 5-ft components.

2. Click on the **From Scratch** tool. Enter the 5-ft numeral in the **Measurements Input Box** and press **Enter** (**Fig. 9.21**).

Figure 9.21 Grid components spacing entered in the **Measurement Input Box.**

Click to select an initial starting point, and drag the mouse cursor across the desired axis, in the **Measurements Input Box**, enter 100 ft and press **Enter.** Repeat the same procedure for the other side of the grid. The completed grid should have a final look as shown in **Fig. 9.22**.

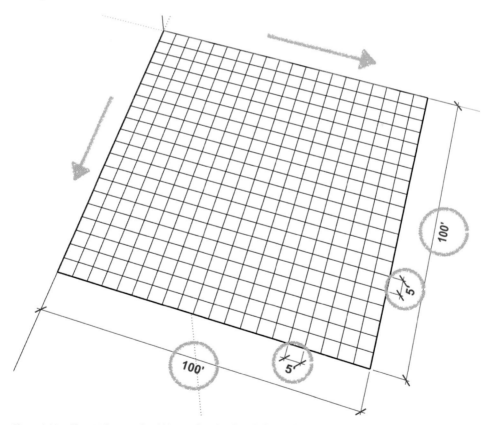

Figure 9.22   The 100-ft × 100-ft grid is completed and ready for further modeling.

Modeling Tip

The grid snaps automatically to the distance that was specified in the **Measurements Input Box**, at the beginning of the operation—this is the 5-ft × 5-ft distance specified in step 2. If you notice, as shown in **Fig. 9.23**, during the grid creation, that the **Measurements Input Box** indicates irregular numbers as the mouse cursor moves away from the starting point. This is nothing to be alarmed by; the completed grid will be spaced as it was specified in the **Measurements Input Box** (5 ft × 5 ft), and will not include any irregular numbers.

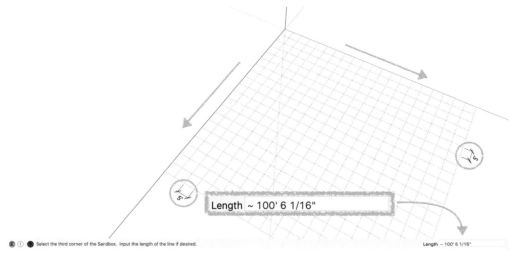

Figure 9.23   Irregular number appears in the **Measurement Input Box** during the grid creation process.

Modeling Tip

The completed grid in step 2 is grouped by default. This is to prevent any possible "stickiness" action between the grid and other components in your model. A word of advice: never explode the grid, but rather work within the group; this will save you a lot of modeling headaches down the road.

3. The next step is to create the topographic features of the terrain model. Select the **Smoove** tool. The default radius of influence for the **Smoove** tool is 5 ft, which represents the base of a topographic feature, that is, an elevation or depression type. For this modeling purpose, change the default setting by entering 20 ft in the **Measurements Input Box** and then click **Enter**. Double-click on the grouped grid, and click on the upper-right corner of the map. Extrude a topographic feature with a 5-ft elevation (**Fig. 9.24**).

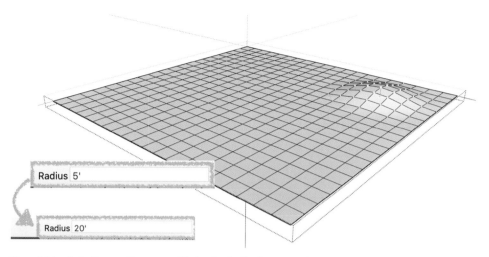

Figure 9.24   Default versus the new specified radius for the **Smoove** tool.

Modeling Tip

When working with the **Smoove** tool, notice the yellow markers positioned at the grid intersections. The yellow markers are more vivid in the center of the radius of influence, and they become less vivid as they move farther away (**Fig. 9.25**). This signifies at what rate the grid surfaces will be modified when extruded upward or downward.

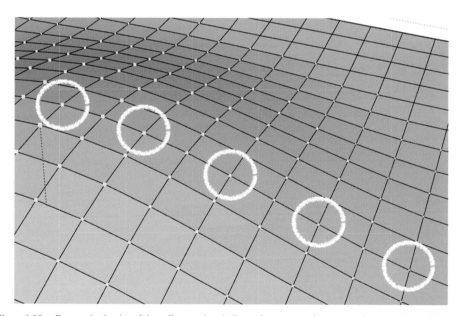

Figure 9.25   Changes in the size of the yellow markers indicate the amount of movement between the grid boxes.

4. Aside from the radius of influence option, topographic features can be created by preselecting a set of grid components. Double-click on the grouped grid and select individual grid components as shown in **Fig. 9.26a**.

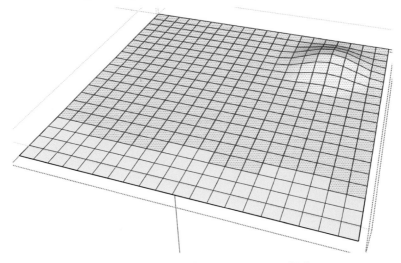

Figure 9.26a   Selecting specific grid sections in order to create a topographic feature.

Click and move your mouse to select large areas of the grid, as necessary. When the selection is complete, click on the **Smoove** tool and then click over the selected grid components. Extrude the topographic feature to a 5-ft elevation (**Fig. 9.26b**).

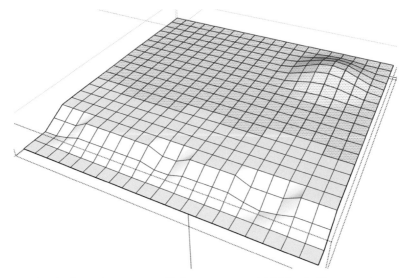

Figure 9.26b   The first stair-step topographic feature was created. Additional grid components were selected to create a secondary, stair-step topographic feature.

Repeat the same step one more time, but offset your selection 20 ft (four grid components back) each time. The final result should look similar to **Fig. 9.26c**.

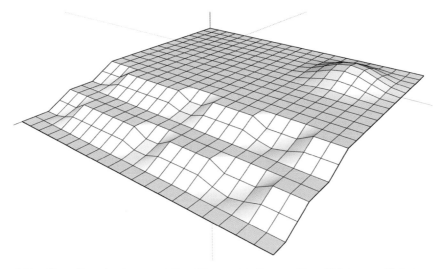

Figure 9.26c   Three-dimensional representation of the completed terrain with multiple topographic features.

Modeling Tip

The example shown in step 4 is particularly beneficial when working with data from a field survey or a topographical map. For example, after the grid is created, select areas of the grid per the contour lines and extrude each level following the elevation information shown in the survey data or the topographic map.

Modeling Tip

When utilizing the **Smoove** tool, keep in mind to change the radius of influence in order to create finer details.

---

5. In the event that you are dissatisfied with the size of your initial grid, and there is a need for adding additional detail or accuracy to the terrain model, the **Add Detail** tool can be utilized. The first step is to access the terrain group, by a double mouse click. Select the area of the grid where you would like to add extra detail (**Fig. 9.27a**).

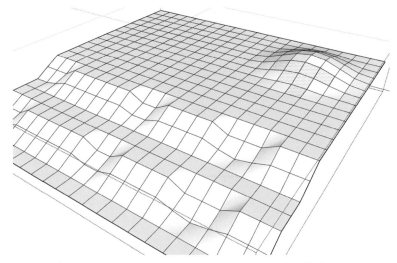

Figure 9.27a   Selecting a section of grid where extra detail is required to be added.

When the selection is complete, click on the **Add Detail** tool; automatically the area of the grid that was previously selected will be divided into smaller sections (**Fig. 9.27b**). Repeat the outline process in steps 3 and 4 to continue with the terrain modeling work.

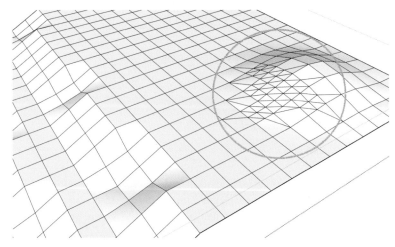

Figure 9.27b   The **Add Detail** tool will further subdivide the selected grid in order to make it easier to add more elevation detail where needed.

Modeling Tip

When working with finer subdivisions of the grid, as you created in step 5, keep in mind to also reduce the radius of influence on the **Smoove** tool. You will get more accurate results this way.

6. The last step in the terrain modeling exercise is to make it a more cohesive element. In the current state, the entire terrain model is composed of a multitude of individual surfaces that are interlinked together (**Fig. 9.28a**).

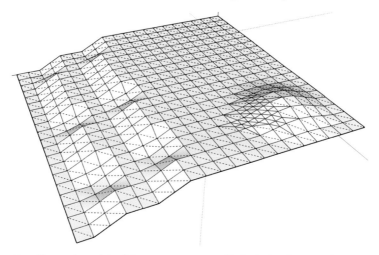

Figure 9.28a   The terrain model and the support geometry of individual surfaces, interlinked together.

By accessing the terrain group (click anywhere on it) you will most likely select one or a group of surfaces. For the purpose of making the grid more cohesive, use the **Soften/ Smooth Edges** dialog box. In order to access the dialog box, from the main menu, select **Window** and then click on the **Soften Edges** option—the **Soften/Smooth Edges** dialog box will appear (**Fig. 9.28b**).

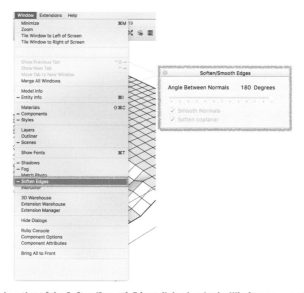

Figure 9.28b   Location of the **Soften/Smooth Edges** dialog box in the **Window** menu and an overview of the **Soften/Smooth Edges** dialog box.

In order to smooth the edges, the first step is to select the terrain group, adjust the **Angle Between Normals** slide bar, and click the **Smooth Normals** and **Soften Coplanar** options. The terrain model will become a cohesive surface (**Fig. 9.28c**).

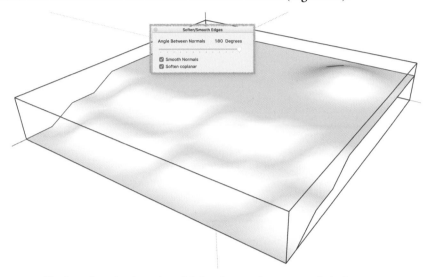

Figure 9.28c   The three-dimensional terrain model after the smoothness was applied to it.

Step 6 concludes the exercise of creating a terrain model utilizing the tools from the **Sandbox** toolset. This option of making terrain models is a great deal more involved than simply importing geo-location data, as you did in the first example. The positive side of this method is that it gives you more latitude and freedom in your initial modeling work. Also, it gives you the ability to make multiple iterations or changes to the terrain model, that is, because of design changes, construction progress, and so on.

The next section of this chapter concentrates on the ways that you can use the created terrain models in your everyday workflow.

Utilization of Terrain Models

The last section of this chapter is a review through an exercise of the different ways you can use the terrain models that were created in the previous two sections. For the purpose of this exercise, take a look at two very helpful tools, the **Stamp** tool and the **Drape** tool. These two tools are part of the **Sandbox** toolset. Similar to the last exercise, the model will be created freehand.

The exercise is separated into two parts. The first part concentrates on the review of the necessary workflow to create a three-dimensional terrain, and the second part of the exercise concentrates on locating and adding a foundation model.

1. The first step is to create a three-dimensional representation of the terrain. You can use any data from your past projects, or you can try to reproduce the terrain shown in **Fig. 9.29**. Use the same method outlined in the **Three-Dimensional Terrain Modeling** section. For this exercise purpose create a 150-ft × 150-ft grid, utilizing the **From Scratch** tool, with 5-ft × 5-ft grid surface increments **Fig. 9.30**.

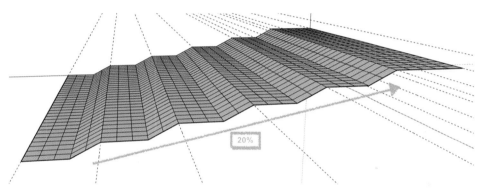

Figure 9.29   Three-dimensional representation of the terrain and the terrain incline.

Modeling Tip

Figure 9.29 reveals that the terrain has a 20%, stair-step type of increase in elevation, which means that for every 25 ft of horizontal run the terrain experiences 5 ft of vertical rise. In order to reproduce this topographical feature, a series of assist lines can be used as elevation and length guides. Step 2 shows a working example, utilizing the assist line method.

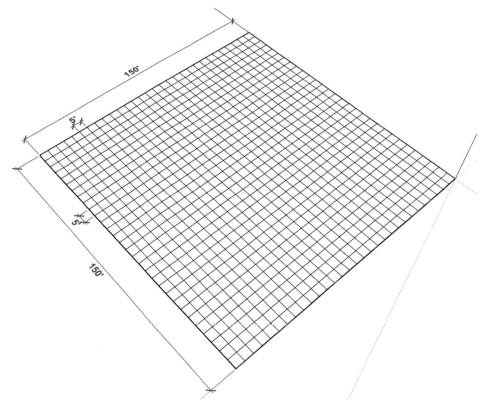

Figure 9.30   Creation of the necessary 150-ft × 150-ft grid for the terrain example.

2. After the grid is created, turn your attention to the assist lines. Based on the examples elevation data, you will have to create a total of 6 assist lines, spaced at 5 ft on-center, on the **Blue** axis. For the horizontal assist lines, each line will be based on its vertical counterpart. Starting from the front of the grid, place the first horizontal assist line at elevation 0 ft and a horizontal distance of 150 ft. The second horizontal assist line will be placed at an elevation of +5 ft and a horizontal distance of 125 ft; repeat this process for all other assist lines. At the end the grid and assist lines should look similar to **Fig. 9.31**.

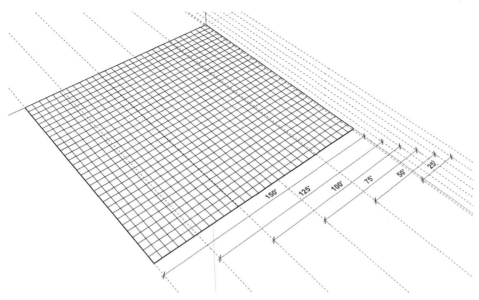

Figure 9.31   Addition of assist lines in order to mark the elevation changes in the terrain model.

3. The next step is to create the topographic feature. Double-click on the grid group and select all the terrain surfaces, except the last batch that is marked by the assist line (**Fig. 9.32**).

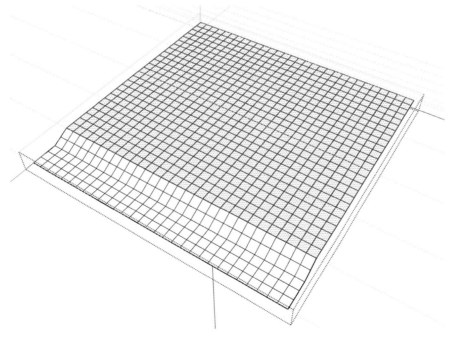

Figure 9.32   Creation of the first stair-step geographic feature.

Select the **Smoove** tool, and enter a radius of influence of 20 ft in the **Measurements Input Box**. With the **Smoove** tool, click on the center of the grid, and elevate all the selected grid surfaces for a total of 5 ft. You will notice that the elevated surfaces intersect the first assist lines. Repeat the same process, but remember to offset your surface selections by 5 ft every time. The final look of the terrain model should look similar to **Fig. 9.33**.

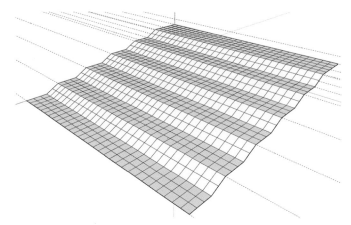

Figure 9.33   Final view of the completed stair step terrain.

4. The last step is to soften the edges. Select the grid by left-clicking on it and utilizing the **Soften/Smooth Edges** dialog box. Increase the **Angle Between Normals** value by sliding the selection bar; furthermore, select the **Smooth Normals** and **Soften Coplanar** options. The grid will be transformed into one cohesive surface (**Fig. 9.34**).

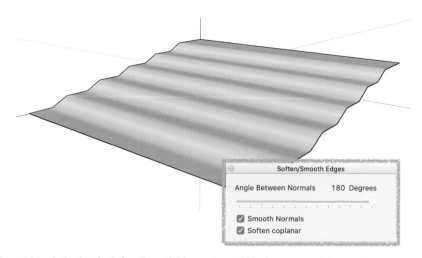

Figure 9.34   Activating the **Soften/Smooth Edge** options to hide the excess model geometry.

5. With the three-dimensional terrain surface completed, turn your attention to locate the bridge abutment footings. For example purposes, use the foundation plan shown in **Fig. 9.35**. Reproduce the foundation plan to the side of the three-dimensional terrain surface.

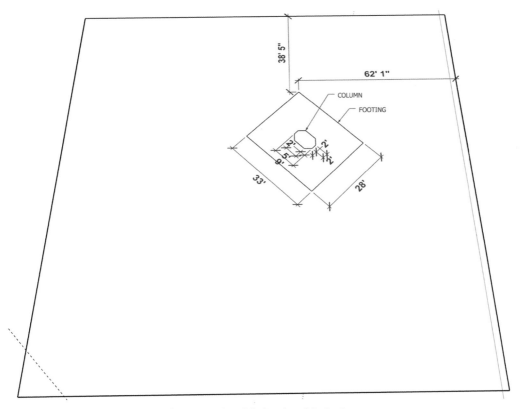

Figure 9.35   Two-dimensional representation of the location of the footing.

Modeling Tip

For actual work applications, import the project foundation plan including the stationing line into SketchUp, following what you learned in **Chapter 3—Importing Files, AutoCAD, and SketchUp**. By doing so, the foundation plan and the stationing line can be traced versus reproduced from scratch. This will save you valuable modeling time and allow you to accurately place the foundation based on the survey data. Always remember to verify the dimensions when you import a file into SketchUp to avoid further modeling issues.

6. The next step is to use the **Drape** tool in order to reproduce the foundation plan content onto the three-dimensional terrain model. Since the terrain grid was created from the origin point of the axis, use this point to align your foundation plan. Select the foundation plan, and click on the corner.

Modeling Tip

For actual work applications, the foundation plan will be located based on the terrain model (stationing information)—primarily by using the intersection between the same coordinate point on the station line that is part of the foundation model and the station line that was created during the terrain modeling.

7. Select the foundation plan again, and offset it in the vertical direction (**Blue** axis), slightly above the terrain model (**Fig. 9.36**).

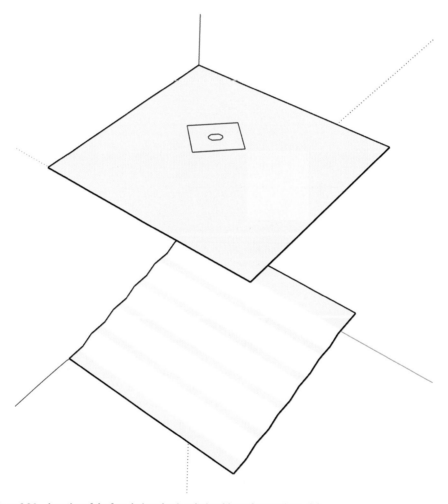

Figure 9.36   Location of the foundation plan in relationship to the terrain model.

8. While the foundation plan is still selected, click on the **Drape** tool. SketchUp will request you to click on the grid that you want to use to drape the foundation plan onto. Click on the terrain model (grid), at which point SketchUp will drape the entire content of the foundation plan onto it (**Fig. 9.37**).

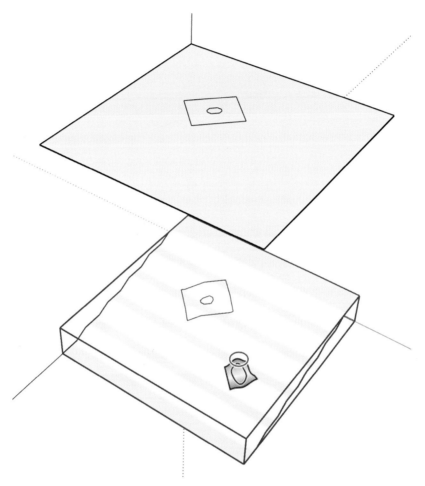

Figure 9.37   Imprint of the foundation plan on the terrain model. The imprint was created using the Drape tool.

Although the terrain model is grouped, SketchUp will modify the group and overlay the foundation plan directly onto the terrain model surface. If you enter the group by double-clicking on it, you will notice that the terrain surface is broken into separate parts—per the foundation plan drawing (**Fig. 9.38**).

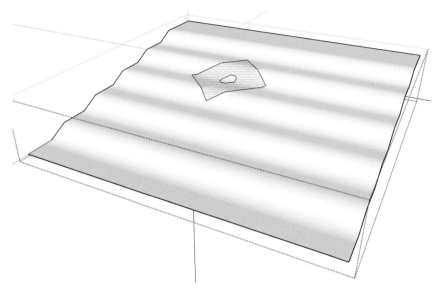

Figure 9.38   Location where the terrain model surface was broken by the imprint of the footing.

9. The final step in this exercise is to locate the correct bottom elevation of the footing and to place the premodeled footing components. For our exercise purpose, offset the bottom elevation of the footing 5 ft from the top of the terrain. Select the **Tape Measure** tool, click on the corner on the **Green** axis and move vertically up, by locking the **Blue** axis (press the up key to lock the axis), until it meets the highest point of the footing. From the initial assist line, place a secondary assist line, 5 ft below the initial one (**Fig. 9.39**). Double-click on the terrain model and delete the internal surface of the footing.

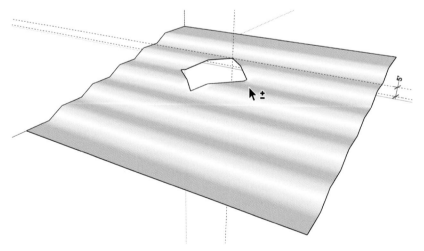

Figure 9.39   Assist lines are used to estimate the depth of the footing. For real-life applications, the depth of the footing can be estimated based on the survey data and specific points on the terrain.

From the **Components** dialog box, choose your footing component and place your footings into the model, using the guide point as an elevation marker (**Fig. 9.40**). In addition, you can add other details such as color, vegetation, equipment, people, and so on.

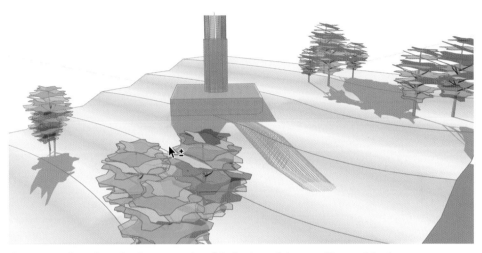

Figure 9.40   **Three-dimensional representation of the footing and the general layout of the site.**

The preceding exercise steps are a great way to visually communicate the location of an already-constructed bridge footing. The question arises, What are your options to visually communicate the construction phase of the footing and the sloping of the soil around it? For this purpose use the **Stamp** tool, which is also part of the **Sandbox** toolset. The following workflow reviews a series of steps that will guide you on how to achieve this visual representation. For the exercise, use the same footing sizes and the same terrain model (cohesive surface with softened edges) that you created in the previous example.

1. The first step in the workflow is to account for the thickness of the formwork panels and any type of ground support those panels will need (**Fig. 9.41**). For modeling purposes, assume that the formwork thickness will be 1 ft and that the ground supports for the formwork will be placed at a 45-degree angle, which will give a total distance of 6 ft (tan(45) $\times$ depth of the footing (6 ft) = 6 ft). An additional 4 ft will be added to account for any work area needed, which makes the total distance 10 ft (**Fig. 9.41**).

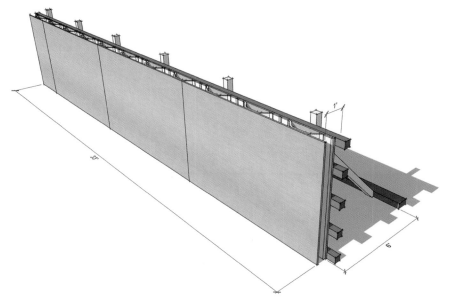

Figure 9.41 Three-dimensional representation of the formwork that will be used for the construction of the foundation.

2. Add the offset distance calculated in step 1 to the overall footing cross-section. Utilizing the **Rectangle** tool, create an equivalent geometrical surface of the footing plus the offset distance to the side of the terrain model (**Fig. 9.42**).

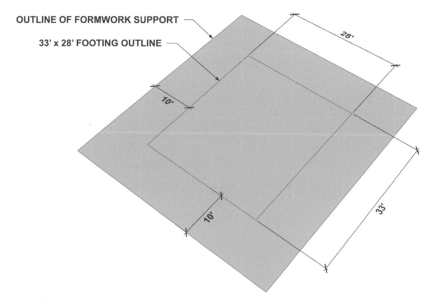

OUTLINE OF FORMWORK SUPPORT

33' x 28' FOOTING OUTLINE

28'

10'

10'

33'

Figure 9.42 The outline of the footing in addition to the outline of the formwork supports.

3. Select the geometrical surface of the footing plus the offset distance and place it above the terrain model; in order to place it correctly, use the footing plan and assist lines as a guide (**Fig. 9.43**). Since this is an example, place the footing at any location you wish.

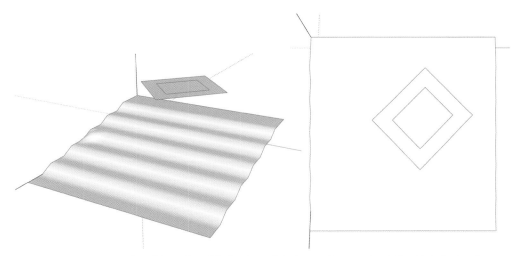

Figure 9.43   Location of the outline of the footing and the formwork support in relationship to the terrain.

4. The next step is to give an offset mark for the bottom of the footing hole. For example purposes, make the total depth 10 ft from the highest contour elevation. This can be completed by utilizing the **Tape Measure** tool and adding assist lines as shown in **Fig. 9.44.**

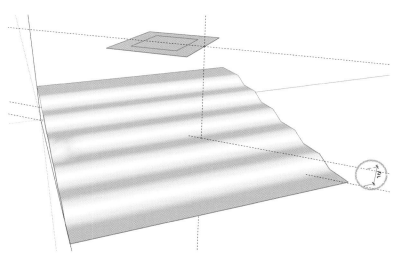

Figure 9.44   Assist lines were created to mark the depth of the footing excavation. On construction projects, the surveyor will give you elevation marks on the terrain for reference purposes.

5. Select the geometrical surface of the footing plus the offset distance and click on the **Stamp** tool. The **Stamp** tool will ask you to enter an offset distance. The offset represents what distance you want to slope the walls of the soil. For example purposes, the sloped distance will be calculated at a 45-degree angle; therefore, based on the height of the soil, which is 20 ft, enter 20 ft (**Fig. 9.45**).

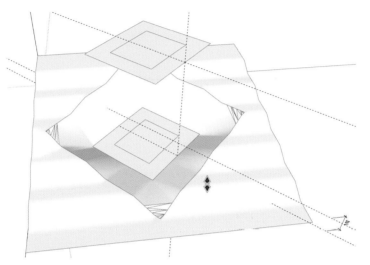

Figure 9.45   The **Stamp** tool was used to create an excavation area into the terrain model. Based on our calculations, the excavation area has the proper soil slopes.

6. Select and hide any unnecessary geometry that was created from step 5. This can be accomplished by selecting the necessary geometry, and from the context menu select **Hide** (**Fig. 9.46**).

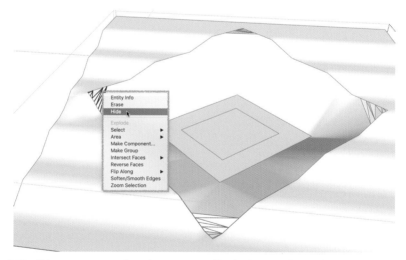

Figure 9.46   Hide any unnecessary lines that were created by the **Stamp** tool using the **Hide** option from the context menu.

7. The foundation excavation is completed (**Fig. 9.47**)!

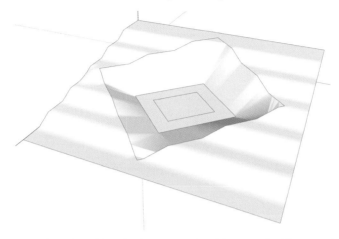

Figure 9.47   The completed three-dimensional representation of the excavation. The excavation also shows the outline of the footing for reference purposes.

8. The last step is to add all the necessary accessories. Following the steps from the previous workflow example, use the **Drape** tool to overlay the contour lines. From the **Components** dialog box, choose your formwork components, reinforcement bars, construction equipment, and anything else you want to show in the work plan. Place all these components in the correct location utilizing the draped outline as markers (**Fig. 9.48**). Do not forget to use what you learned in **Chapter 4— Introduction to Information Modeling and Organization** with regard to organization when adding components to your mode.

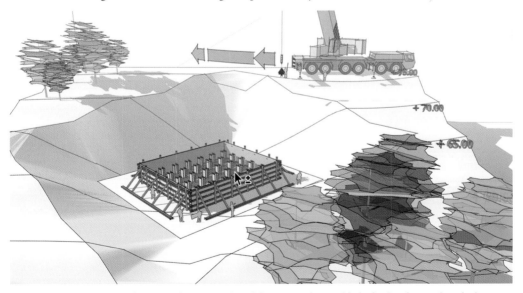

Figure 9.48   Three-dimensional representation of the excavated area with the footing, formwork, and other accessories. This type of presentation is useful during work plan reviews.

The last tool in the **Sandbox** toolset that is reviewed is the **Flip Edge** tool. From personal experience, use the **Flip Edge** tool when you want to fill in a valley type of geometry that was created intentionally or unintentionally on the terrain model (**Fig. 9.49**).

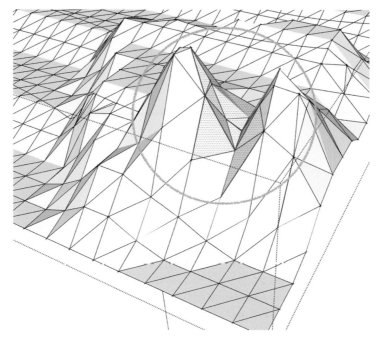

Figure 9.49   Individual terrain surfaces are selected around a topographic feature that requires fill.

In order to fill in the geometry, the first step is to reverse the smoothness of the terrain model from the **Soften/Smooth Edges** dialog box. Select the terrain and de-select the **Smooth Normals** and **Soften Coplanar** options, and also slide the selection bar for the **Angles Between Normals** option back to zero. Select the **Flip Edge** tool, and click on grid lines on the outside and inside of the valley, as shown in **Fig. 9.50**; SketchUp will

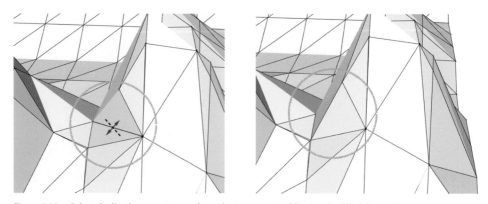

Figure 9.50   Select the line between two surfaces that you want to fill using the **Flip Edge** tool.

automatically fill in the section that was selected with the **Flip Edge** tool. Continue select-ing until the valley is filled in. One thing to remember is that utilizing the **Flip Edge** tool can drastically alter the overall geometry of your terrain model. Be cautious and understand the consequences when using this great modeling tool.

This concludes the chapter on three-dimensional terrain modeling. My sincere hope is that this chapter was above all fun to do and insightful and that you learned something new in terrain and foundation modeling that you can utilize on current or future project.

CHAPTER 10

Modeling of Construction Mechanization

Modeling of Construction Equipment and Range of Movement

Planning tips for equipment modeling

Dynamic Components and Construction Equipment

Conclusion

Projects rely heavily on a large variety of construction mechanization in order to complete different aspects of the work. This chapter is designed to work through a series of steps needed to detail an accurate representation of construction mechanization, with the ultimate purpose of being used in three-dimensional modeling. In addition, dynamic components and how you can incorporate them into your mechanization models to reproduce the range of movements are discussed. There are two key concepts to remember when you are at the starting point of construction mechanization modeling, and those are **review** and **balance (Fig. 10.1)**.

Figure 10.1   Few examples of different types of cranes.

Review—The first priority is to review the specification sheets and understand the machine in question. Using inaccurate machine properties such as location of the rotation pin, boom pin, width of axles, location of jacks, length of boom, and so on can create enormous problems with serious consequences if the model is used for work planning purposes.

Balance—Any mechanization model should have the right balance between visual accuracy and the actual size of the three-dimensional model. Find your own personal style of how your models are represented and make a determination of what is enough detail for you. You can create the most accurate three-dimensional model of a machine, but what good will that do if you cannot use it because it increases the demand on the computer system and slows down the response time of the SketchUp application. **Figure 10.2** shows an example of my personal style when reproducing construction mechanization in three dimensions.

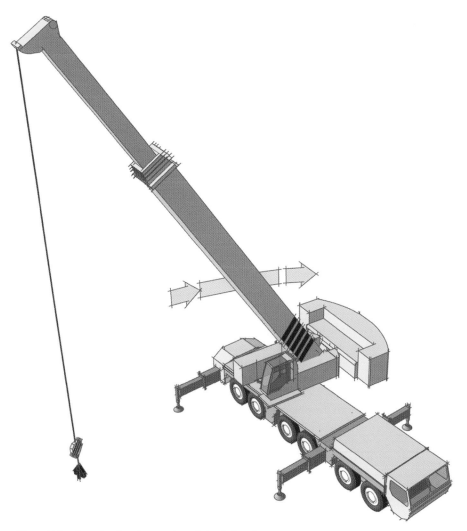

Figure 10.2   Reproduction of a crane in a minimalistic style.

Modeling of Construction Equipment and Range of Movement

This section reviews the important steps when modeling a three-dimensional representation of equipment. The exercise reviews an all-terrain crane. For those who are less familiar with this type of crane, the all-terrain crane is a mix of truck-mounted and rough terrain crane. Basically, the all-terrain crane has the ability to travel on public roads and go over uneven construction terrain with its agility and maneuverability (**Fig. 10.3**).

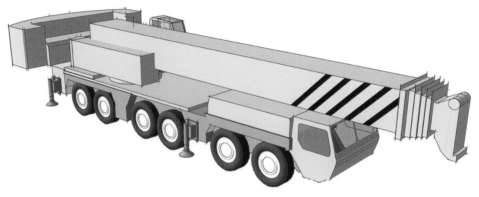

Figure 10.3   All-terrain crane.

As mentioned previously, each modeler should create their own style of how to represent the construction equipment and at the same time create a balance between visual representation (the amount of detail used in the model) and the overall equipment accuracy. The following workflow was designed to show the main characteristics of creating a three-dimensional model of some equipment, in this case the all-terrain crane. This will be a review of the key elements in the three-dimensional model of the all-terrain crane.

- **Keep It Simple but Accurate**—Always keep your equipment visually simple. Any equipment, with enough time and patience, can be reproduced to the smallest detail. Although it sounds tempting, there is no reason for it. Most of the equipment used in three-dimensional models specifically created for construction projects is there to explain a certain work activity that will be completed on the project. In comparison to the object that is either being lifted or worked on, the equipment is fairly small, and all that extra detail will be lost in the background. Certain parts of a piece of equipment will have to be accurately reproduced, and those are the overall width, length, and height; location of the rotating pin; location of the outriggers (if you are reproducing a truck-type crane); swing radius; counterweight dimensions; boom dimensions; cab location; and so on. All these parts have a direct impact on the accuracy of the work plan that you plan to portray and later present in front of your team. Anything else, such as motor, exhaust ports, toolboxes, interior equipment, and the like should be omitted, because they are only visual enhancers and at the same time computer resource drainers. **Figure 10.4** shows the crane and the rotating cab (the boom was omitted for visual purposes). The main parts of the crane are accurately reproduced, and all the extras are omitted.

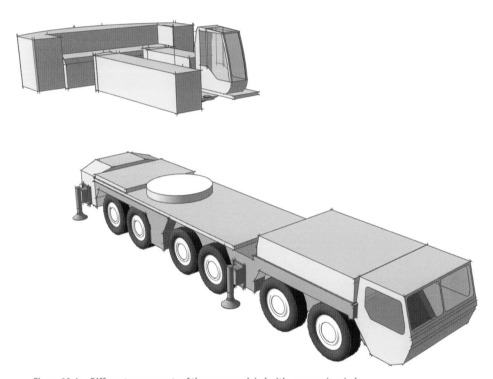

Figure 10.4   Different components of the crane modeled with accuracy in mind.

- **Group the Main Components of a Machine**—Always group the main components that you plan to make dynamic; for example, the main boom of an all-terrain crane consists of multiple boom sections that are extended or retracted based on the operating radius. The main boom would be one of these main components. **Figure 10.5** shows an **X-Ray** view of the main boom with all the different boom sections.

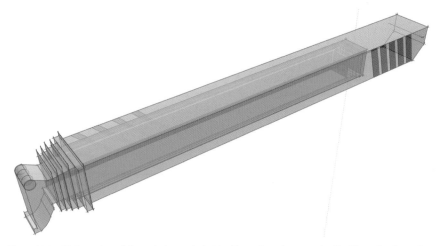

Figure 10.5   X-Ray view of the main boom. Individual boom lengths are tucked inside each other as it would be the case on a real crane.

Each of these boom sections is converted into individual components for ease of usability. For crawler-type crane, where the boom is modular, it is a good idea to have a small library with the available boom sections that you can later combine based on the requirements for a specific job (**Fig. 10.6**).

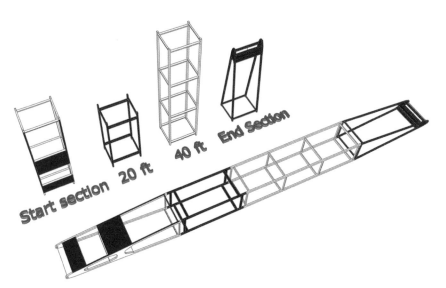

Figure 10.6   Different boom sections for lattice boom crane.

- **Add Hand Movement Locations**—Although rotating components will be converted into dynamic components, it is good to have a secondary option to move them. **Figure 10.7** shows the addition of "handles" to the boom and the operator's platform.

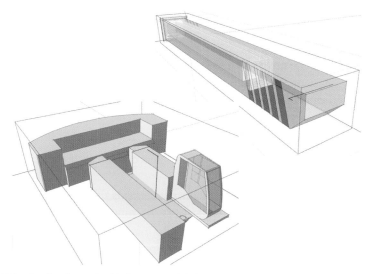

Figure 10.7   Small and unnoticeable lines are modeled in with the components in order to help when rotating the equipment to the correct positions.

- **Create Visual Cues**—The addition of visual cues is a great way to bring attention to a certain danger that the machine can pose to field personal or to show an exclusion zone for the equipment. This can be highly useful when planning work activities and reviewing the safety aspects of the work. One example is the counterweight swing area of the crane or the operational area for a crane. **Figure 10.8** shows these two visual cues.

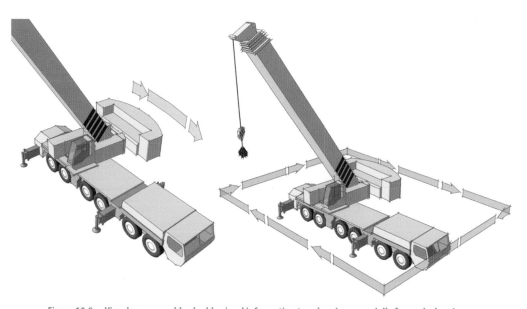

Figure 10.8   Visual cues can add valuable visual information to a drawing, especially for work planning purposes.

Dynamic Components and Construction Equipment

Dynamic components give you the ability to bring accurate movement to static three-dimensional components. This section reviews the necessary steps to create a dynamic component. The workflow example uses the same crane from the previous section. You are more than welcome to create a machine of your own. The process for creating a dynamic component is as follows:

- Create a regular component of each section—if not previously done.
- Create an overall component that will encompass all the other components that you would like to move together.
- Assign movement attributes to the components that you want to make dynamic. Use the **Component Attributes** inspection window for this task.
- Set up how you want to interact with the dynamic portion of the component, again use the **Component Attributes** inspection window for this task.

1. The first step is to determine the type of movements you will need for your mechanized equipment. For example purposes, use the all-terrain crane from the first section of this chapter. Now add two rotational movements, one for the swing mechanism and the other for the crane boom (**Fig. 10.9**). Logically seen, these two

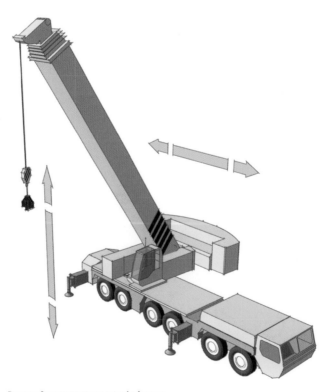

Figure 10.9   Range of movements on a typical crane.

movements represent the most critical movements in the machine from which you can adjust the pick radius and therefore the angle of the boom.

2. From the main menu, click on **Window** and select the **Component Attributes** option, which will open the **Component Attributes** inspection window (**Fig. 10.10**).

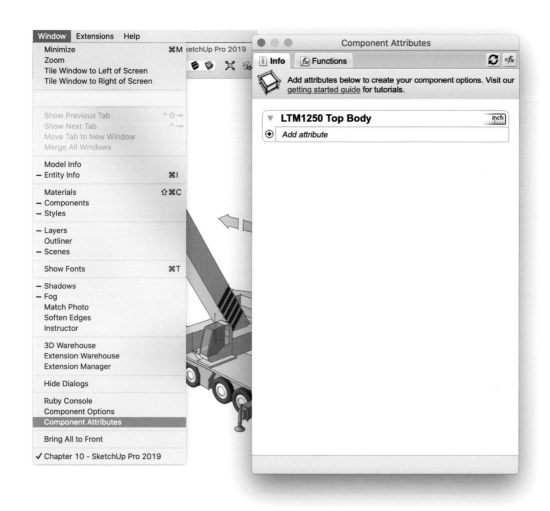

Figure 10.10   The location of the **Component Attributes** inspection window.

3. Select the **Boom** and **Swing Mechanism** by clicking on each component with the left mouse button.

4. Click anywhere on the two selected components with the right mouse button, and from the context menu select the **Make Component...** option, which will bring the **Make Component** inspection window (**Fig. 10.11**).

5. In the **Definition** and **Description** boxes, write **Upper Crane Body** (**Fig. 10.11**).

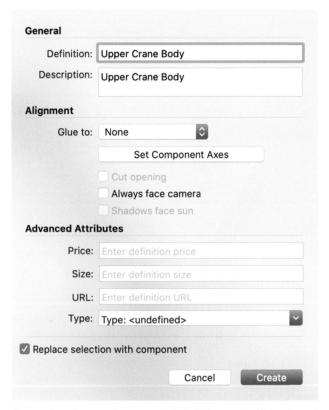

Figure 10.11   Process of creating a single component that will be converted into a dynamic component for movement purposes.

6. Click on the **Set Component Axes** button. The mouse cursor will change from an arrow to the axis symbol. Set the axis cursor in the center of the **Swing Mechanism** and press the mouse button three times in order to set the *x, y,* and *z* axis, respectively **(Fig. 10.12)**.

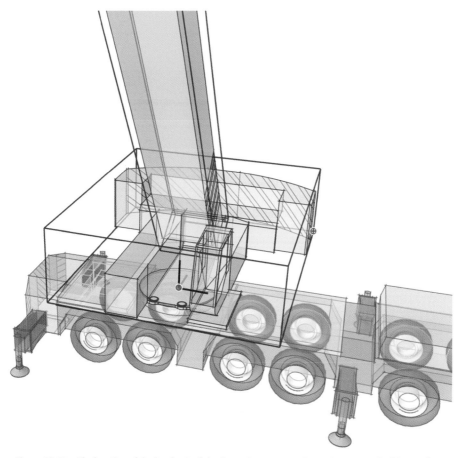

Figure 10.12   The location of the local axis of the dynamic component is very important. In this case the upper body of the crane rotates about a point called a "pin."

Modeling Tip

It is very important to note that in each of the components that will be converted into dynamic components, the local axis has to be set in relationship to the movement you want to achieve and at the center point. For example, for the Top Body component of the crane, the local axis has to be set at the center of the rotation pin with the **Blue** axis in the vertical position. If the local axis is not set properly, the dynamic component will not function properly.

7. Click on the **Create** button and the new **Upper Crane Body** component will be created. Select the newly created component and shift your attention toward the **Component Attributes** inspection window, where you can see all three components: the main or overall component called **Upper Crane Body** and the two secondary components called **Main Boom** and **Swing Mechanism (Top Body)** (**Fig. 10.13**).

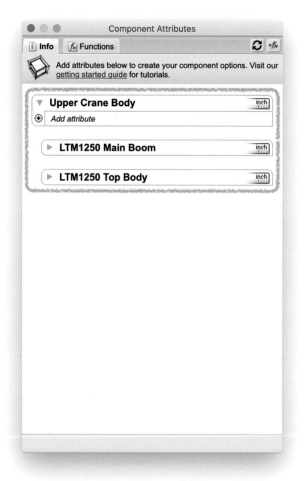

Figure 10.13   The **Component Attributes** inspection window showing the two secondary components (LTM 1250 Main Boom and LTM 1250 Top Body) being part of the main **Upper Crane Body** component.

Modeling Tip

The reason you created an overall component is that when you move the **Upper Crane Body** component you also want the **Boom** and **Swing Mechanism** to move with it.

8. The next step is to assign dynamic movements to the components. Select the **Upper Crane Body** component and click on the **Add Attribute** option, which will open a context menu **(Fig. 10.14)**.

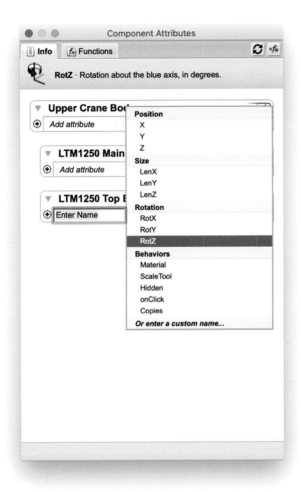

Figure 10.14   Selecting the type of movement from the attributes list.

9. From the context menu select the **RotZ** option that is located under the **Rotation** heading and is colored in blue **(Fig. 10.15)**. Basically you are telling the component to rotate about the vertical axis. If for some reason your model is using a different setup for axes, for example, the **Blue** axis is no longer a vertical but rather a horizontal axis, choose the correct corresponding axis.

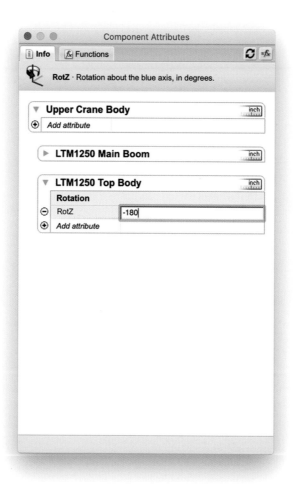

Figure 10.15   The selected attribute will allow the user to rotate the **Upper Crane Body** component around the **Blue** axis of rotation.

10. Click on the **Details** button (menu/arrow), and from the **Display Rule** drop-down menu, select **User can edit as a textbox** option, and press the **Apply** button **(Fig. 10.16)**. Selecting this option will allow you to change the rotation by clicking on the text box, which you can see in step 11.

Figure 10.16   Adding an option on how to interact with the created attribute. In the above shown case, SketchUp will allow you to enter a specific degree numeral for the movement amount.

11. Pressing the **Apply** button will take you back to the **Component Attributes** inspection window. If you double-click with the left mouse button on the "0" number, the text box will allow you to change the angle, which in turn will rotate the **Upper Crane Body** component.

Modeling Tip

The angle option is not additive in nature; rather, it will rotate the structure according to the number typed based on the original position. If you enter 0 (zero) in the text box, the structure will return to the original position.

12. Follow the same exact procedure as outlined in steps 8 through 11 in order to assign a dynamic movement to the Boom component, with one major difference, choose the RotY rotation option (**Fig. 10.17**).

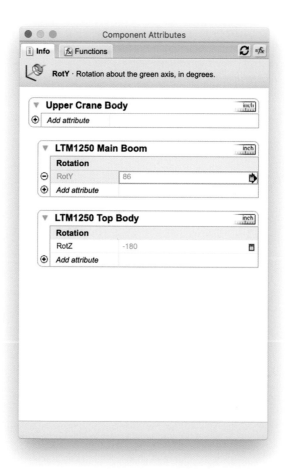

Figure 10.17   The dynamic component is completed for the **Upper Crane Body**. The **Upper Crane** Body can move about the **Blue** axis (for the top body portion) and about the Green axis (for the boom).

Step 12 concludes the creation of the dynamic component. In order to use the dynamic components, enter a certain degree into either component text box. The component will automatically rotate that certain degree. This particular setup of a dynamic component is extremely helpful during girder erection work planning, since the boom location in the crane specification charts is shown in degrees.

I do encourage you to add additional dynamic components on your own. For example, the outriggers of the crane are created to extend outward to certain widths prior to the use of the crane. One option is to add a dynamic component where each outrigger can be extended based on a specific length. Since there are several width options, this will be helpful and save time in the future if you use the same crane at different outrigger widths. In order to do this, you will have to use the **Position** attributes. Similar dynamic components can be added for the **Boom** section too.

Explore the endless possibilities and foremost have fun!

Conclusion

In closing Section 2, as with everything in life, there will be times when the three-dimensional model will not cooperate with you at all, and will probably seem like it has a mind of its own. Take a break, get yourself a cup of Earl Grey hot tea (sci-fi fans will understand this joke) and get back to three-dimensional modeling. There is no reason not to have fun and enjoyable modeling time with SketchUp!!

CHAPTER 11

Styles Inspection Window

Select Submenu
Edit Submenu
   Edge
   Face

Background
Watermark
Modeling
Mix Submenu

The **Styles** inspection window, allows you to either use default or create new visual styles for your three-dimensional models. The type of style you choose, or create, says a lot for you as a designer and also has a profound impact on how the drawing is perceived and comprehended by the end user. This is the gateway where you can leave your design and aesthetic mark on what you create and how you show it to the world.

A very important point to note regarding to the **Styles** dialog box is that changes made to the overall style of the drawing will not impact the actual geometry of your three-dimensional model. Rather, it will modify the visual aspect of edges, faces, shadows, background setting, and the like. SketchUp offers three options on how to discover your personal and unique style, and those are through the use of the preset styles that come standard with the application, creating your own style or a mix and match of options one and two. Before discussing these options, we start with an initial overview of the **Styles** inspection window.

The **Styles** inspection window can be accessed from the main menu by clicking on Window and then selecting **Styles** (**Fig. 11.1**). The **Styles** inspection window can be separated into an upper and a lower portion (**Fig. 11.1**).

Figure 11.1   How to access the **Styles** inspection window and the overall layout of **Styles.**

The upper portion of the **Styles** dialog box will have the options and controls shown in **Fig. 11.2**:

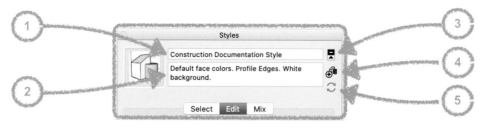

Figure 11.2   The upper portion of **Styles** inspection window with the corresponding controls and options.

1. Provides a name for a specific style.
2. Provides a description of a specific style.
3. Option to display a secondary selection panel.
4. Option to create a new style type.
5. Updates all the changes that were made to a specific style. Furthermore, style changes can also be updated by clicking the update symbol presented on the style icon (**Fig. 11.3**).

Figure 11.3   Locations where the style can be updated after revisions are made or a brand new style is created.

In the lower portion of the **Styles** window, you will find the heart of the **Styles** dialog box, which are the **Select**, **Edit**, and **Mix** submenus. These three submenus are reviewed in more detail in the next section of this chapter (**Fig. 11.4**).

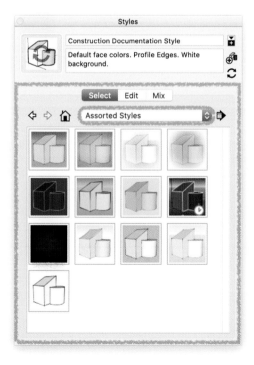

Figure 11.4   Lower portion of the **Styles** window.

Select Submenu

The **Select** submenu consists of a library of different types of styles that are ready to be used just by clicking on them. There is a wide variety of styles, which can give a different look to your model. The drop-down menu, located beneath the three main submenus, will give you a list of all the styles available by assortment (**Fig. 11.5**).

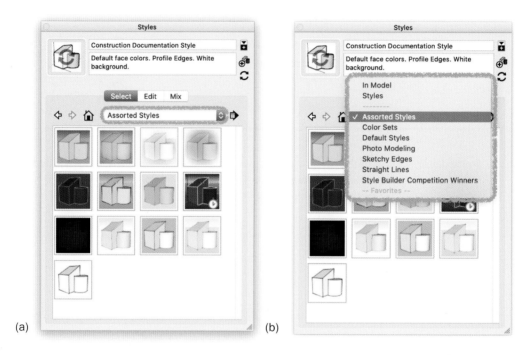

(a)   (b)

Figure 11.5   (a) The available **styles** under the Select submenu. (b) The drop-down menu and the list of **styles** available as part of the Select submenu.

To the left of the drop-down menu, you will find the **Details Arrow**, which will provide you with a context menu and options connected directly to the style collections and also how to visually see the assortment of styles (**Fig. 11.6**). The context menu is self-explanatory.

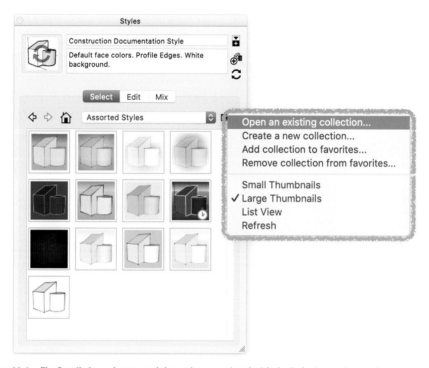

Figure 11.6   The Details Arrow button and the options associated with the Styles inspection window.

Edit Submenu

The **Edit** submenu of the **Styles** dialog box allows you to tailor different options, that is, for default styles or newly created ones. The **Edit** submenu consists of five different areas/ options of available changes: **Edge**, **Face**, **Background**, **Watermark**, and **Modeling**. The five options are reviewed separately, since they will give the most visual impact to your three-dimensional model.

Edge

As the name implies the **Edge** dialog box is tailored toward changes made to the overall line style of your three-dimensional model (**Fig. 11.7**).

Figure 11.7   The **Edit** dialog box and the available options.

The **Edit** dialog box can be separated into three different sections. The first section allows you to choose and apply changes to **Edges** or **Back Edges**. From a personal standpoint, I never make any visual modifications to the **Back Edges** since they are not visible by default, unless there is some type of specific detail you want the viewer to notice.

The second section allows you to make different types of changes, which can have a direct impact on your model. Since these are visual changes and this is a construction handbook, a three-dimensional representation of a brick is used to show the differences.

- The **Profiles** option is selected by default and its main function is to give depth to the profile edges of your three-dimensional model. **Figure 11.8** shows the difference between a standard profile value and an increased profile value.

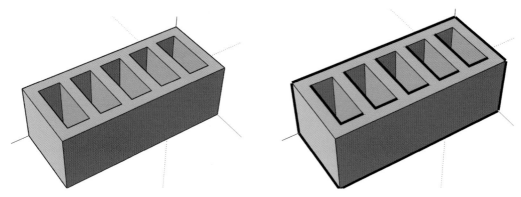

Figure 11.8   The visual difference that the **Profiles** option can create. The brick model to the left has a standard profile value, the brick model to the right has an increased profile value.

- If activated, the **Depth Cue** option creates a sense of depth to your object. The closer your object is toward your line of sight the thicker the lines will appear and vice versa, as you can see in **Fig. 11.9**.

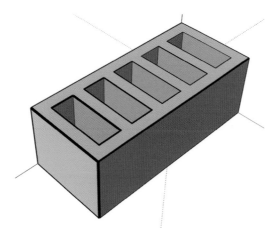

Figure 11.9   The **Depth Cue** adds thicker lines closer to the point of view. Great way to add accent to three-dimensional drawings based on line of sight.

- As the name implies, the **Extensions** option adds extended lines at the end of line intersections. The greater the extensions value is set, the longer the lines will appear (**Fig. 11.10**). One caution: when applying this option to your drawing, although it can provide for an interesting style for your three-dimensional drawing, it can also cause confusion because lines will appear longer than they should be. Use this style with caution in the heavy civil industry, especially when creating work plans.

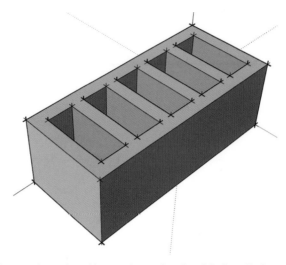

Figure 11.10   The Extension option adds extensions to the edge of the lines. Use it sperinly as it creates an ilusion that lines are longer then required.

- The **Endpoints** will apply corner markers to your three-dimensional model (**Fig. 11.11**). The corner markings can be a great enhancer in areas where you want to accentuate the openings in the structure.

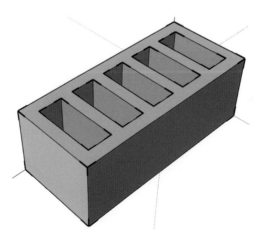

Figure 11.11   The Extension option makes the corners more noticeable.

- The **Jitter** and **Dashes** options come hand in hand. The **Jitter** option provides a sketch-type visual effect to your three-dimensional model (**Fig. 11.12**). From a visual aspect, it can create some very interesting outcomes if paired with other edge options. The visual option should not be used when creating work plans that require precision in the presentation. The **Dashes** option is selected by default and enhances the **Jitter** visual effect.

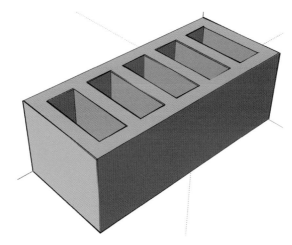

Figure 11.12 **The Jitter and Dashes** provide a smudged look to the three-dimensional model. Similarly to the Extension option, use this option sperinly as it creates an ilusion that the three-dimensional object had double lines.

The last section in the **Edge** dialog box will give you the ability to change the color to the lines on your three-dimensional model. SketchUp does not have the ability to change the color of individual lines as other CAD applications do, but it does give you the ability to change the color by material or by axis.

Face

The **Face** dialog box allows you to make changes to how the surfaces are visually comprehended (**Fig. 11.13**). In the first section of the **Face** dialog box, you can change the standard color scheme of how the front and back surfaces are shown in your model. These colors only apply to surfaces prior to color application. The second portion of the **Face** dialog box allows you to change the overall view of your surfaces. These options were discussed in detail in **Chapter 1** under the Styles toolbox review.

Figure 11.13   The **Face** dialog box and the available options.

Background

The **Background** dialog box gives you the ability to change the background color of your modeling area or to add the sky or ground to your model (**Fig. 11.14**). If you do choose to add the sky or ground to your model, you can also change its color based on your preferences and ultimately the location of your three-dimensional model.

Figure 11.14   The Background dialog box and the available options. The Background dialog box allows you to chnage setting connected to the overall environment where the three-dimensional model is located.

Watermark

The **Watermark** dialog box (**Fig. 11.15**) allows you to add an image to the viewing area of your model. This image can signify your own logo, your company's logo, stage in construction (preliminary, for construction, etc.), or just an image that you wish to apply.

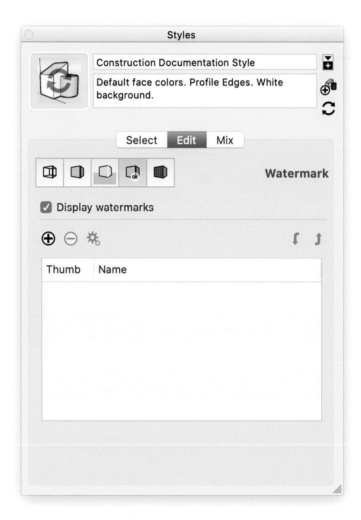

Figure 11.15   The **Watermark** dialog box and the available options. Watermark is a good option to protect your work from unauthorized use or to add a company a name.

When you are ready to add a watermark to your drawing, SketchUp will present you with multiple options in order to achieve your vision for it, from the ability to change or select the location, scale, transparency, color, and so on (**Fig. 11.16**).

Figure 11.16   Available options when creating a watermark.

Modeling

The last dialog box under the **Edit** tab is **Modeling** (**Fig. 11.17**). The **Modeling** dialog box allows you to make changes on how you perceive different tool options by changing their color coding. In addition, you can also activate or deactivate certain view options, for example, hidden geometry, guide lines, axes, and so on.

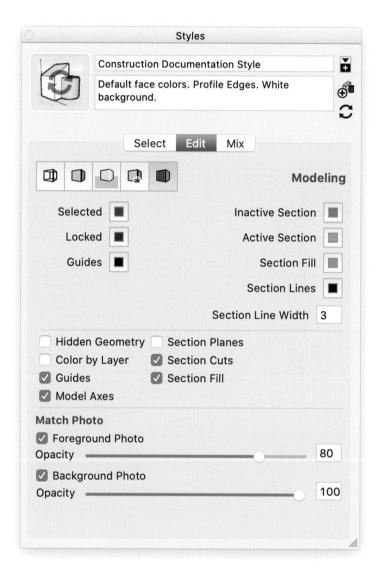

Figure 11.17   The **Modeling** dialog box and the available options. The Modeling dialog box aloows you to chnage how certain options are shown in SketchUp. Pay attention to the Section cut options (right side of the dialog box), since you will use them on a regular basis.

Mix Submenu

The **Mix** submenu allows you to apply certain visual presets to your model by choosing an existing style and dropping it in the corresponding settings bin. The setting from the existing style will only change and update the setting to where it was placed and keep all the other ones unchanged. This is an easy way to create styles without reviewing settings for each style that you like. You can mix and match different settings from different styles and create something unique (**Fig. 11.18**).

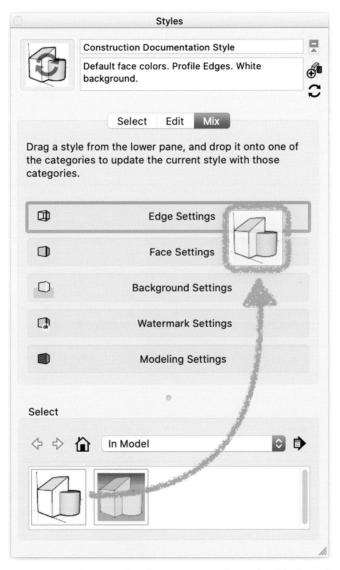

Figure 11.18   Ability to drag and drop a specific style into a corresponding settings bin. Depending which bin you use, thouse certain properties of the style will be updated. A great way to mix and match proprties that you like.

The **Styles** dialog box gives you a unique outlet to develop and practice your own style. The key thing to remember is that a lot of practice and ultimately experimenting with different settings can lead you to develop your own unique style. The ultimate idea is to have fun and see where your style curiosity can take your work.

CHAPTER 12

Introduction to Scenes, Section Cuts, Shadow, and Fog

Scenes	Shadow Tool
Section Cuts	Fog Tool
Section Plane Tool	

This chapter discusses options for visual effects that can be applied to your three-dimensional model. Visual effects play an important factor in how the three-dimensional model is presented and ultimately comprehended by the end user. An accurate three-dimensional model goes hand in hand with a stunning visual presentation that will explain the concept and at the same time engage the viewer.

For the purpose of visual effects, four different tools are reviewed: **Scenes, Section Cuts, Shadows,** and **Fog**.

Scenes

SketchUp allows you to save specific scenes of your model that can be used later in your modeling work for further visual improvements or to export into LayOut. This was slightly touched on in the **Scenes** dialog box in **Chapter 3—Importing Files, AutoCAD, and SketchUp**, but it will be highly beneficial to expand on some of the additional options present in the **Scenes** dialog box.

The **Scenes** dialog box can be activated by clicking on the **Window** menu and selecting the **Scenes** option. The **Scenes** dialog box can be separated into two sections. Section one displays the standard set of options, and section two displays all the additional options that are active by default (**Fig. 12.1**).

Figure 12.1   The location of the **Scenes** window and the corresponding options available.

The first set of scene options are as follows:

1. The **Update** icon allows you to update/save any changes to all or a specific scene.

2. The **Plus** sign (+) allows you to add a new scene.

3. The **Minus** sign (−) allows you to remove one or multiple scenes.

4. The **Up** and **Down** arrows allow you to position the scenes in a specific order of your choosing. Changes in the scene order will also impact the scene buttons that are generated as one is created. The Scenes buttons can be found below the menu bar.

5. The **View Option** icon allows you to set your preferred way of how you view the scenes in the **Scenes** dialog box.

6. The **Show Details** icon allows you to toggle the additional option. This is discussed in more detail later in this section.

7. The **Details Arrow** represents the last icon in the series of the standard set of options. When selected, the details arrow will open a context menu that will have the same option as the six options discussed earlier.

The second set of **Scenes** options are activated by default and can be deactivated based on your personal set of preferences or concept you have for that specific scene.

1. **Name and Description**—This is self-explanatory; this is the location where you name the scene and add a specific description that will further explain the scene.

2. **Include in Animation**—SketchUp will add an animation transition as you shift between scenes.

3. **Camera Location**—The camera location option will save the visual properties of the scenes; basically, it make a note on the zoom distance, point of view, and so on.

4. **Hidden Geometry**—Any geometry that is deemed hidden will stay hidden when the scene is activated.

5. **Visible Layers**—Makes a note of which layers are visibly active and which are not and therefore will only show the active visible layers.

6. **Active Section Planes**—Saves the cut section properties for that specific scene.

7. **Style and Fog**—Saves the related properties to a specific style and fog settings for the scene.

8. **Shadow Settings**—Saves all the shadow-related properties, from time of day, month, or year, for that scenes.

9. **Axes Location**—The last option directly affects if the axes display in the scene and their relative position.

Scenes can be of great benefit to a modeler, especially for presentation purposes where you can shift between views and setting with one click of your mouse. Think of scenes as a snapshot in time of your entire model or sections with specific settings attached to them. Similar to the other visual effect tools reviewed in this chapter, practice is the best way to get familiar with the properties and how they affect your model. Similar to the **Scenes** dialog box, scenes only affect the visual portion of the model and not the actual model geometry. Deleting a scene does not mean deleting a portion of your model; it only removes that particular snapshot in time with the preselected options.

Section Cuts

The utilization of section cuts represents a good way to show a specific detail that is located internally in your model, without the need to disassemble it. The sections toolset consists of four different tools: **Section Display**, **Section Cuts**, **Section Fill**, and **Section Plane Tool** (**Fig. 12.2**).

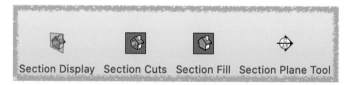

Figure 12.2   Icon representation of the section cut line of tools.

The **Section Plane Tool** represents the actual tool that creates a section cut in your model and the rest of the section tools represent visual options that can be applied to the section cuts in your three-dimensional model. Before reviewing the necessary steps to create a section cut, it is important to review the visual options first.

Section Display tool—**By** toggling the **Section Display** tool on and off you are toggling the visual representation of the location of the section cut. Hiding the visibility of the visual representation of your section cut does not have an effect on section cut itself (**Fig. 12.3**).

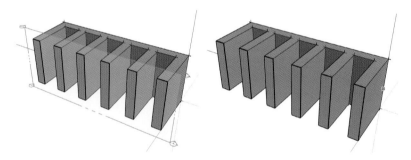

Figure 12.3   Visual difference when the **Section Display** tool is activated and deactivated.

Section Cuts tool—**By** toggling the **Section Cuts** tool, you activate or deactivate the section cut.

Section Fill tool—The **Section Fill** tool adds a surface to all the components that are cut by the section plane (**Fig. 12.4**).

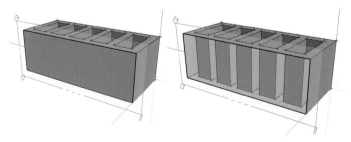

Figure 12.4   The visual difference when a section fill is applied to a section cut. The color and look of the section fill can be modified in the Styles inspection window.

The color and visual aspect of this surface can be modified in the **Styles** dialog box, under the **Modeling** option (**Fig. 12.5**).

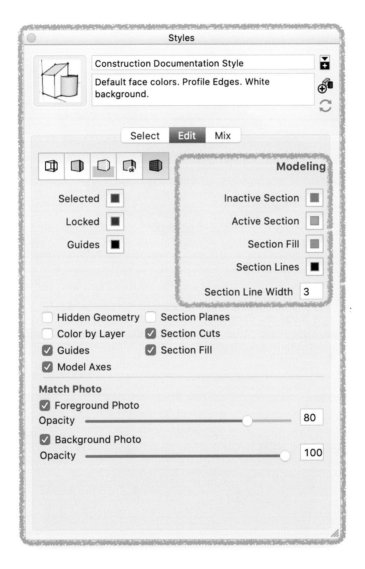

Figure 12.5   Available options in the Styles inspection window to change different section cut options, including the section fill color.

Section Plane Tool

Think of section cuts as regular surfaces, not any different from when you create a rectangle with the **Rectangle** tool; therefore, the same general properties of movement, rotation, and snap also apply to section cuts. The first and the most important step in creating a section cut is to visualize what you intend to show and how you intend to show it.

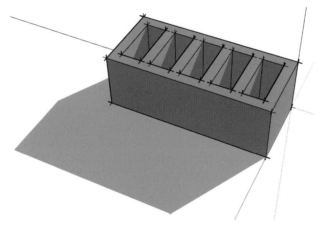

For the purpose of explaining section cuts, use the trusty brick.

In order to create a section cut, click on the **Section Plane** tool, and then select a face of the three-dimensional model that you wish to assign as the starting point of your cut. The next step is to assign a **Name** and **Symbol** for your section cut, which can be done from the **Name Section Plane** dialog box. After you click on the **OK** button, SketchUp will cut the three-dimensional model (**Fig. 12.6**).

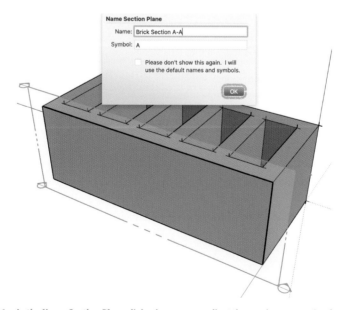

Figure 12.6   In the **Name Section Plane** dialog box you can adjust the naming convention for your section cuts.

The section cut can be moved to the final position by utilizing the **Move** tool in the same fashion as you would move a regular surface. Similar to regular surfaces, the snap properties also apply to section cuts:

- **Up Arrow** key will restrain the section cut movement to the **Blue** axis.
- **Left Arrow** key will restrain the section cut movement to the Green axis.
- **Right Arrow** key will restrain the section cut movement to the **Red** axis.
- **Down Arrow** key will restrain the section cut movement to a specific place, and the section cut will be represented in a Magenta color.

Although section cuts are usually applied to a specific face of a three-dimensional model, there will be instances where you will want to place the section cut on an angle to a plane. The steps necessary to accomplish this task are simple. If you recall from the beginning of this section, section cuts are similar to surfaces; therefore, the same general properties apply as they do for surfaces. In order to place a section cut on an angle, your first step is to place the section cut on the surface perpendicular to the angle you want to achieve with the section cut. The next step, after assigning a **Name** and **Symbol** to the section cut, is to select the **Rotate** tool. With the **Rotate** tool selected, click in the center or any other portion of the section cut, and rotate the section cut until you achieve the desired angle. The last step is to move the angled section cut to the final position on the three-dimensional model, utilizing the **Move** tool (**Fig. 12.7**).

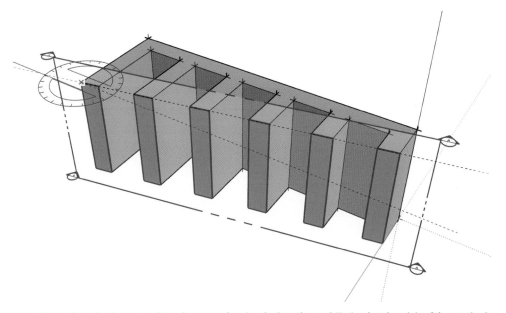

Figure 12.7   Section cut positioned on an angle using the **Rotation** tool. Notice that the origin of the rotation is located at the intersection point of the brick edge.

Only one section cut can be active at a time. Although this is the rule, there will be times when you will need multiple section cuts in order to show a specific detail of your model. Multiple active section cuts can be added as long as they are located in a different group of components in the overall model. Each separate component can have an active section cut at the same time. Back to the trusty brick to explain the process. The brick is turned into a component, and as such you will be able to assign two separate section cuts at the same time. If you want to add two different section cuts to the brick, the first step is to apply a section cut as you did at the beginning of the chapter. The next step is to double-click on the brick component, in order to access the geometry, and apply a secondary section cut (**Fig. 12.8**).

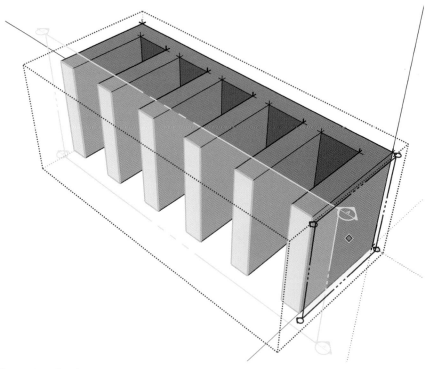

Figure 12.8 Creating a secondary active section cut by activating the component. Notice the brick component is active in order to add a secondary section cut.

When you exit the component, the three-dimensional model will have two active section cuts at the same time. Always remember to group or to make different parts of your model into components (**Fig. 12.9**).

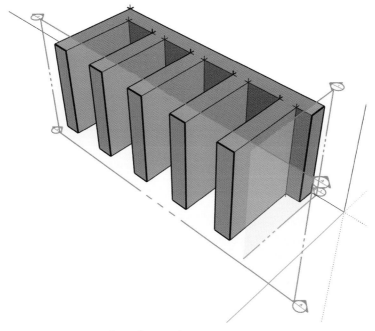

Figure 12.9   Two section cuts active at the same time.

The last section that is covered with regard to section cuts is the color coding for each section cut.

- **Orange Section Cut**—Represents an active section cut Fig. 12.10(a).

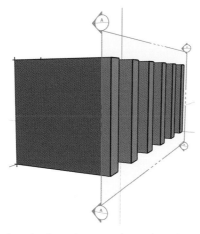

Figure 12.10(a)   Orange color code of a section cut - active section cut.

- **Blue Section Cut**—When the symbol of the section cut is filled, the section cut is selected and active. If the section cut symbol appears not filled, the section cut is not activated Fig. 12.10(b).

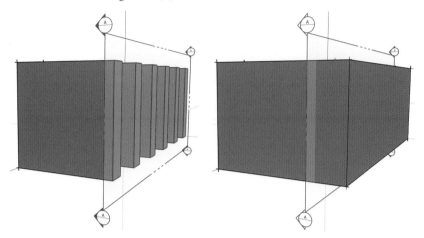

Figure 12.10(b)   Blue color code of a section cut. The section cut to the left is active and selected, the section cut to the right is not active.

- **Grey Section Cut**—Represents an inactive section cut Fig. 12.10(c).

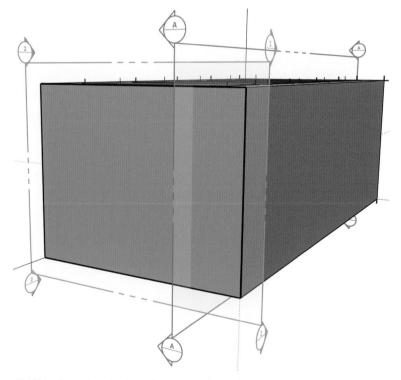

Figure 12.10(c)   Grey color code of a section cut - inactive section cut.

Shadow Tool

The **Shadow** tool represents a great way to add an emphasis of reality to your three-dimensional models. The shadow is generated from each individual object in your model. With that fact in mind, use the shadow effect only after you have completed all the three-dimensional modeling and you are working on the visual effects portion. The reason for this suggestion is the fact that SketchUp will refresh the shadows each time you update your model geometry or change the angle of view. The constant refreshing of the shadow increases the demand on your computer and thus decreases the response time of the application.

Besides a realistic emphasis, the **Shadow** tool can also be used to create a sun study on your three-dimensional model, since it generates the shadows based on location, the month of the year, and the time of day.

There are two options to activate the **Shadow Settings** dialog box. The first option is to select the **Shadows** tool from the toolbar. The second option is to click on the **Window** menu and select **Shadows** (**Fig. 12.11**).

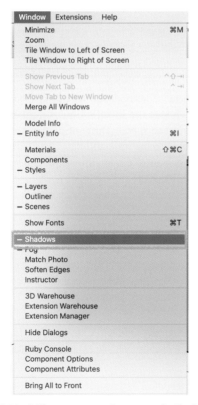
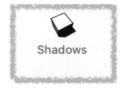

Figure 12.11   Different ways to gain access to the **Shadow Settings** dialog box.

Either way you choose, SketchUp will open the **Shadow Settings** dialog box (**Fig. 12.12**).

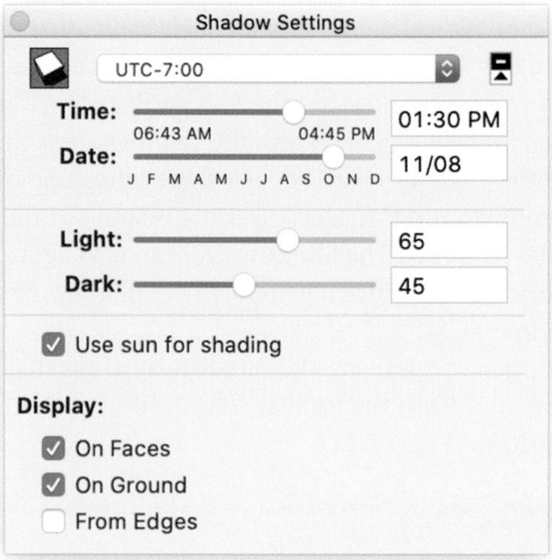

Figure 12.12   The **Shadow Settings** dialog box.

The **Shadow Settings** dialog box can be separated into three main sections as shown in **Fig. 12.13**. In the first section you can choose the specific time zone that your project is located in. In addition, you can also change the specific time of day and month for which SketchUp will base the shadows. The changes to the time and date can be done in two ways, either by using the sliding bar or, for more accurate control, by manually inputting them into the input boxes located next to the sliding bar.

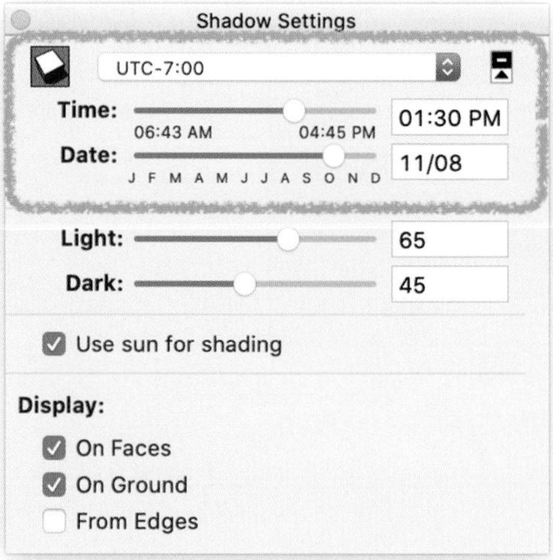

Figure 12.13   The upper portion of the **Shadow Settings** dialog box and the available options.

The second section of the **Shadow Settings** dialog box offers the user the option to make changes with regard to the amount of light and darkness that will be present in the drawing. This is personal taste and also depends on the context of the drawing; in summary it all depends on what you want to portray to the end user. The amount of light and darkness can be adjusted by using the sliding bar or, for more accurate control, by manually inputting them into the input boxes next to the sliding bars (**Fig. 12.14**).

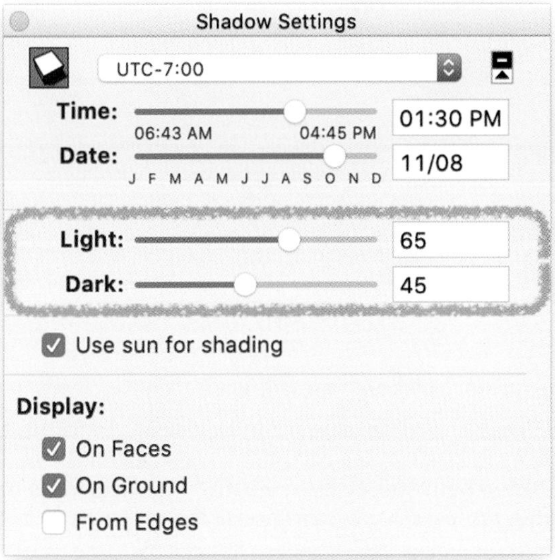

Figure 12.14   The middle portion of the **Shadow Settings** dialog box and the available options.

The last section of the **Shadow Settings** allows you to set how the shadows are displayed in your model. These settings are self-explanatory and therefore will not be explained further (Fig. 12.15). If you choose the **Use sun for shading** option when the **Shadow** tool is activated, your three-dimensional model will generate shadows, and at the same time the entire model will become more vivid.

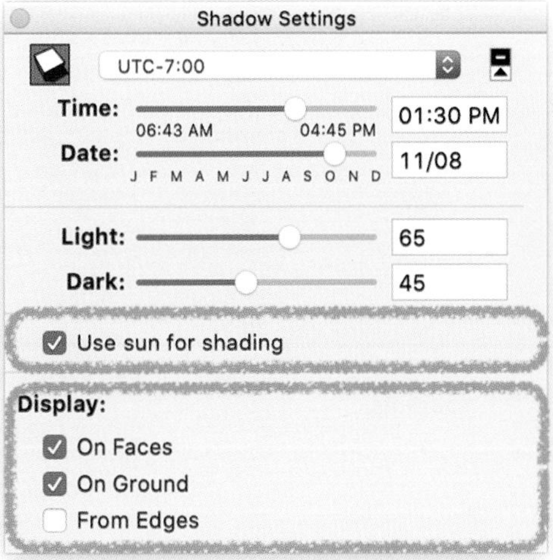

Figure 12.15   The lower portion of the **Shadow Settings** dialog box and the available options.

Colors become more realistic and crisp see Fig. 12.16(a) and 12.16(b) for difference in color when the shadow option is not activated and activated.

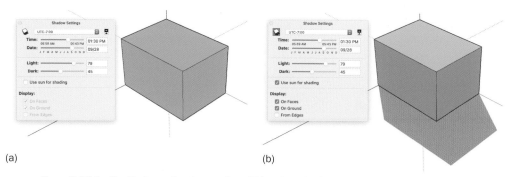

(a)

(b)

Figure 12.16(a)   The Shadow options is not active, which makes colors less vivid and chrisp. (b) The Shadow option is active, which makes colors vivid and chrisp.

There will be instances where you will want to have clear and realistic display of the colors in your model but without shadows. When the **Use sun for shading** is activated without the **Shadow** tool being active, SketchUp will generate light onto your model, as it is being lighted from a light source but your three-dimensional objects will not generate shadows (**Fig. 12.16(c)**).

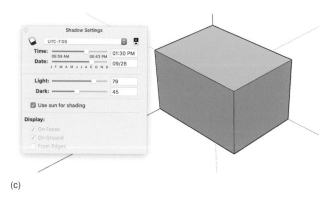

(c)

Figure 12.16(c)   The Shadow option is not active, but the Use Sun for Shading option is, which makes colors vivid and chrisp.

Fog Tool

Fog can be a great way to focus the attention of the end user to a specific section of your three-dimensional drawing. The **Fog** dialog box can be activated by clicking on the **Window** menu and selecting the **Fog** option, at which point the **Fog** dialog box will appear (**Fig. 12.17**).

Figure 12.17   Different ways to get access to the **Fog** dialog box, and the available options.

In order to start the fog option, click on the **Display Fog** check box, located in the upper portion of the **Fog** dialog box. The two sliding bars under the **Distance** section allow you to set the intensity of the fog that will be displayed in your model. The top slider bar sets the starting vantage point for your fog—how far from the datum zero you want the fog wall to be located, with the right side representing 0 and the left side representing infinity. The second sliding bar represents the back wall of the fog. The closer the second slider is advanced to the first slider, the thicker the fog will appear on your three-dimensional model; the opposite is also true. From personal experience, practice is the best way to acquire a "feel" for the visual properties of fog and how they will impact your model (**Fig. 12.18**).

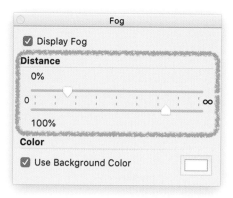

Figure 12.18   The sliding bars under the **Distance** options allow you to tailor the location and intensity of the fog present on your drawings.

The last section of the **Fog** dialog box is **Color**. By default, the background color of your model also represents the color that will be used for your fog. In certain situations where you want to change the default setting of the fog color, click on the color box and select a new fog color (**Fig. 12.19**).

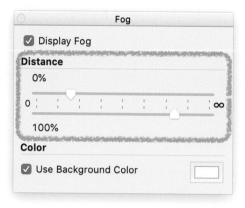

Figure 12.19   The **Color** option allows you to change the color of the fog.

In conclusion to **Chapter 12**, with all the tools that were covered in this chapter it is very important to provide yourself with plenty practice in order to fully understand how these tools function and also how they can positively impact your model. Remember not to get discouraged by "mistakes" because, from personal experience, every mistake I have made by utilizing either an incorrect setting or tool taught me a new trick or a procedure. The most important thing to remember is to always have fun with everything you do!

Index